AF389009

Synthesis and Characterization of New Superconductors Materials

Synthesis and Characterization of New Superconductors Materials

Editors

Edmondo Gilioli
Davide Delmonte

MDPI • Basel • Beijing • Wuhan • Barcelona • Belgrade • Manchester • Tokyo • Cluj • Tianjin

Editors
Edmondo Gilioli
Institute of Materials for Electronics
and Magnetism (IMEM) of the
National Research Council (CNR)
Italy

Davide Delmonte
Institute of Materials for Electronics
and Magnetism (IMEM) of the
National Research Council (CNR)
Italy

Editorial Office
MDPI
St. Alban-Anlage 66
4052 Basel, Switzerland

This is a reprint of articles from the Special Issue published online in the open access journal *Crystals* (ISSN 2073-4352) (available at: https://www.mdpi.com/journal/crystals/special_issues/ Superconductors_Materials).

For citation purposes, cite each article independently as indicated on the article page online and as indicated below:

LastName, A.A.; LastName, B.B.; LastName, C.C. Article Title. *Journal Name* **Year**, *Article Number*, Page Range.

ISBN 978-3-03943-004-8 (Hbk)
ISBN 978-3-03943-005-5 (PDF)

© 2020 by the authors. Articles in this book are Open Access and distributed under the Creative Commons Attribution (CC BY) license, which allows users to download, copy and build upon published articles, as long as the author and publisher are properly credited, which ensures maximum dissemination and a wider impact of our publications.
The book as a whole is distributed by MDPI under the terms and conditions of the Creative Commons license CC BY-NC-ND.

Contents

About the Editors

Edmondo Gilioli is Senior Researcher at IMEM-CNR (the Institute of Material for Electronic and Magnetism of the National Research Council) in Parma, Italy.
His main interests are (i) synthesis (by conventional and high pressure techniques) of metastable bulk materials (mainly superconductors and multiferroics) and (ii) deposition by physical vacuum techniques of films for energy applications, such as thin film solar cells and 2D superconductors (coated conductors). More info can be found on his ORCID profile: http://orcid.org/0000-0001-6973-2179.

Davide Delmonte Ph.D. is a physicist and Postdoctoral Researcher at IMEM-CNR (Institute of Material for Electronic and Magnetism of the National Research Council) in Parma, Italy.
Since 2012, he has been working on unconventional methods to synthesize and characterize different multifunctional inorganic compounds, i.e., high-T superconductive perovskites (cuprate-based), multiferroic perovskites, multiple ordered systems, ferroelectrics, and ferrophotovoltaics. He is author of about 30 peer-reviewed scientific works. More information can be found on his ORCID profile: https://orcid.org/0000-0001-5367-527X.

Editorial

Synthesis and Characterization of New Superconductors Materials

Edmondo Gilioli * and Davide Delmonte *

IMEM-CNR, Institute of Materials for Electronic and Magnetism-National Research Council,
Parco Area delle Scienze 37/A, 43124 Parma, Italy
* Correspondence: edmondo.gilioli@cnr.it (E.G.); davide.delmonte@imem.cnr.it (D.D.)

Received: 28 July 2020; Accepted: 28 July 2020; Published: 29 July 2020

In the last few decades, the persisting scientific interest in high temperature superconductor (HTS) cuprates has been accompanied by the search for new families of superconducting compounds (SPCs). Notable examples are the intermetallic borides (e.g., MgB_2), iron-nickel-based superconductors (e.g., La(Fe,Ni)(Pn,Ch)O, (Pn,Ch = pnictide or chalcogenide ions), heavy fermions (e.g., $CeCoIn_5$), organic (doped-fullerene, e.g., Rb_3C_{60}, or fulvalene-based, e.g., $(BEDT-TTF)_2X$), superhydrides systems (e.g., H_3S), etc., which strongly revitalized the research in this field.

Historically, the synthesis of new SPCs has played a crucial role; with the dream of increasing the critical temperature (Tc), a number of new SPC phases, as well as chemical substitutions or structural modifications induced by external stimuli (such as high pressures or magnetic field) of known compounds have been obtained. This massive effort still leads to important achievements: (i) on a more fundamental level, to refine and sometimes to develop theoretical models to explain the intriguing phenomenon of the superconductivity, (ii) to understand the correlation between structural and transport properties in the matter and (iii) to improve the SPC's properties in view of best performing practical applications.

The Special Issue "Synthesis and Characterization of New Superconductors Materials" of *Crystals* has been conceived to address various issues, providing reviews and insights on the most promising systems, as well as innovative solutions for practical applications.

K. Zhou et al. [1] systematically reviewed the crystal and electronic structure, the chemical substitution and the pressure- and thickness-dependent of FeSe-based SPC. Among the most interesting results reported: (i) the SPC properties vs. chemical substitution on the Fe or Se sites, (ii) the role of the application of high pressure in enhancing Tc and simultaneously induces magnetism, (iii) the organic and inorganic layer intercalated-FeSe leading to a Tc increase from 30 to 46 K and (iv) a further Tc increase due to the electron-phonon coupling between single FeSe unit cell and $SrTiO_3$ substrate.

M. van Delft et al. [2] showed a clear evidence of a two- and three-dimensional superconducting phases in type-I Weyl semi-metals tantalum phosphide (TaP); remarkably, they revealed Tc between 1.7 and 5.3 K in as-grown TaP single crystals at ambient pressure, while all the known reports only found SPC through the surface contact with a sharp tip or the application of high pressure. They also showed how the Weyl semi-metals and other topological phases can host exotic phenomena such as non-zero momentum pairing or the Majorana fermion, a viable candidate for the ultimate realization of a scalable quantum computer.

T. Tani et al. [3] studied the field-angle dependence of interlayer magnetoresistance in crystals of organic conductor $\alpha(BEDT-TTF)_2I_3$ grown by the electrolysis method, under strong magnetic field. The authors demonstrated a qualitative agreement between the theory and experimental results of the field-angle dependence in this system. Besides, the quantum Hall ferromagnetic state assumed in the present study supports the helical surface state.

HTS cuprates with perovskite structure are still the most widely studied SPCs; although the "original" compounds cannot be defined as new materials, a variety of unconventional chemical substitutions or treatments are reported to address either fundamental or practical issues.

Cabassi et al. [4] presented the role of chemical substitutions on (Bi,Pb)-2212 (BSCCO) HTS. They correlated the Tc and the structural morphology of Zn, Y, Ti, and Nd chemically substituted compounds. Significantly, the elements are incorporated in the BSCCO-2212 structure at amounts exceeding the ranges currently found in the literature. In a few samples, the appearance of higher Tc is associated with the formation of BSCCO-2223.

N. Mohd Hapipi et al. [5] reported a comparative study on the AC susceptibility of $YBa_2Cu_3O_{7-\delta}$ (YBCO) added with $BaZrO_3$ (BZO) nanoparticles prepared via solid-state and co-precipitation method. Besides providing an analysis of the correlation between BZO concentration and YBCO grain size, the susceptibility measurement showed a decrease in the Tc onset with BZO addition, attributed to the distribution of BZO particles at the grain boundaries.

Z. Zhang et al. [6] optimized the synthesis of $ErBa_2Cu_3O_{7-\delta}$ (Er-123) to be used as a SPC joint between commercial a $GdBa_2Cu_3O_{7-\delta}$ (Gd-123)-coated conductor. They systematically studied the effect of sintering parameters on the phase formation, microstructure and superconducting properties of Er-123 powder to obtain a high performing SPC joint. This work aims to improve the development of applications requiring long length HTS coated conductors.

A waiting for new exciting discoveries on superconductivity, we hope that the scientific community enjoys reading this Special Issue!

Funding: This research received no external funding.

References

1. Zhou, K.; Wang, J.; Song, Y.; Guo, L.; Guo, J. Highly-Tunable Crystal Structure and Physical Properties in FeSe-Based Superconductors. *Crystals* **2019**, *9*, 560. [CrossRef]
2. Van Delft, M.; Pezzini, S.; König, M.; Tinnemans, P.; Hussey, N.; Wiedmann, S. Two- and Three-Dimensional Superconducting Phases in the Weyl Semimetal TaP at Ambient Pressure. *Crystals* **2020**, *10*, 288. [CrossRef]
3. Tani, T.; Tajima, N.; Kobayashi, A. Field-Angle Dependence of Interlayer Magnetoresistance in Organic Dirac Electron System α-(BEDT-TTF)2I3. *Crystals* **2019**, *9*, 212. [CrossRef]
4. Cabassi, R.; Delmonte, D.; Abbas, M.; Abdulridha, A.; Gilioli, E. The Role of Chemical Substitutions on Bi-2212 Superconductors. *Crystals* **2020**, *10*, 462. [CrossRef]
5. Mohd Hapipi, N.; Lim, J.; Chen, S.; Lee, O.; Shaari, A.; Awang Kechik, M.; Lim, K.; Tan, K.; Murakami, M.; Miryala, M. Comparative Study on AC Susceptibility of YBa2Cu3O7−δ Added with BaZrO3 Nanoparticles Prepared via Solid-State and Co-Precipitation Method. *Crystals* **2019**, *9*, 655. [CrossRef]
6. Zhang, Z.; Wang, L.; Liu, J.; Wang, Q. Synthesis of ErBa2Cu3O7−δ Superconductor Solder for the Fabrication of Superconducting Joint between Gdba2cu3o7−δ Coated Conductor. *Crystals* **2019**, *9*, 492. [CrossRef]

© 2020 by the authors. Licensee MDPI, Basel, Switzerland. This article is an open access article distributed under the terms and conditions of the Creative Commons Attribution (CC BY) license (http://creativecommons.org/licenses/by/4.0/).

Article

Field-Angle Dependence of Interlayer Magnetoresistance in Organic Dirac Electron System α-(BEDT-TTF)$_2$I$_3$

Takehiro Tani [1,*], Naoya Tajima [2] and Akito Kobayashi [1]

[1] Department of Physics, Nagoya University, Furo-cho, Chikusa-ku, Nagoya 464-8602, Japan; akito@s.phys.nagoya-u.ac.jp
[2] Department of Physics, Toho University, Funabashi, Chiba 274-8510, Japan; naoya.tajima@sci.toho-u.ac.jp
* Correspondence: tani@s.phys.nagoya-u.ac.jp

Received: 11 March 2019; Accepted: 16 April 2019; Published: 19 April 2019

Abstract: The effect of the Coulomb interaction in interlayer magnetoresistance is elucidated in collaboration with theory and experiments for the Dirac electron system in organic conductor α-(BEDT-TTF)$_2$I$_3$ under a strong magnetic field. It is found that the effective g-factor enhanced by Coulomb interaction depends on the angle of the magnetic field, resulting in the field-angle dependence of a characteristic magnetic field in which interlayer resistance has a minimum due to spin splitting $N = 0$ Landau levels. The qualitative agreement between the theory and experimental results for the field-angle dependence of interlayer magnetoresistance is obtained.

Keywords: Dirac electron; Landau level; interlayer magnetoresistance; organic conductor; α-(BEDT-TTF)$_2$I$_3$

---

1. Introduction

The electron correlation effects in two-dimensional Dirac electron systems have attracted much attention [1–5]. In the Dirac electron system of organic conductor α-(BEDT-TTF)$_2$I$_3$ [6–13], it was shown that the electron correlation effects become a key factor to understanding electronic properties [14–22], since the Coulomb interaction is comparable with the band width. Moreover, the Fermi energy almost coincides with the Dirac point in α-(BEDT-TTF)$_2$I$_3$; thus, any other energy bands do not overlap with Fermi energy [23,24]. α-(BEDT-TTF)$_2$I$_3$ has a clean Dirac electron system, since the density of impurity is estimated to the ppm order [23,24]. These features are also advantageous to developing the physics of the correlated Dirac electron system.

The layered structure of α-(BEDT-TTF)$_2$I$_3$ enables interlayer magnetoresistance measurements, which revealed the anomalous electronic properties of two-dimensional Dirac electron systems connected by weak interlayer tunneling [25–30]. In the magnetic field normal to the conductive layer, the energy of Landau levels in a massless Dirac electron system is expressed as $E_N = \pm\sqrt{2e\hbar v_{\mathrm{F}}^2 |N| B}$, where $\hbar$, v_{F}, N, and B denote the Planck constant, the Fermi velocity of the Dirac cone, the Landau index, and the magnetic-field strength, respectively. One of the characteristic features in these systems is the appearance of $N = 0$ Landau levels at the Dirac points. This effect was detected in interlayer magnetoresistance under a transverse magnetic field [25,26]. Interlayer magnetoresistance primarily depends on the interlayer tunneling of the Landau carriers, where Landau carriers indicate the carriers belonging to Landau levels that contribute to electric current. Note that in each Landau levels there are states with density proportional to B. Thus, the magnetic field creates mobile $N = 0$ Landau level carriers. The effect of the magnetic field appears only through the change of the $N = 0$ Landau level carrier density at the vicinity of the Dirac points. Thus, negative interlayer magnetoresistance due to the increase of the degeneracy of the $N = 0$ Landau levels was observed [25,26]. It was also shown that interlayer resistance has a minimum for $g\mu_B B_0/2 \cong \hbar/\tilde{\tau}$ due to the Zeeman splitting of the $N = 0$ Landau levels, where g, μ_B, and $\tilde{\tau}$ denote the g-factor, the Bohr magneton, and the relaxation time, respectively. If the electron correlation effects in the Landau levels are negligible, B_0 is independent of the angle of the magnetic field.

The electron correlation effects in the Landau levels, however, have been controversial. Although the possible ordered states due to Coulomb interaction, such as the valley-ordered state [31–33] and the interlayer spin-ordered state [34,35], were proposed, it was suggested that the anomalous increase of the spin lattice relaxation rate at low temperatures [33] can be explained by the spin transverse fluctuation in the absence of ordered states [36].

In the present study, we investigate the effects of the electron correlation between $N = 0$ Landau level carriers on interlayer magnetoresistance as a function of field-angle θ from the interlayer axis in collaboration with theory and experiment. The effective g-factor, g^*, is treated using the mean field theory of the Coulomb interaction between the tilted Dirac electrons in the quantum limit. It is numerically shown that effective Coulomb interaction V_{HS}, which enhances g^*, is approximately proportional to $(B\cos\theta)^\gamma$, where γ depends on the tilt of the Dirac cone. It is found that a characteristic magnetic field B_0, at which the interlayer resistance has a minimum, depends on θ and temperature T, where the inverse of B_0 is proportional to $\cos\theta$ approximately and the coefficient increases as T decreases. These results are in qualitative agreement with the experiment.

2. Method

2.1. Formulation

In α-(BEDT-TTF)$_2$I$_3$, there are two band-crossing points called Dirac points being assigned as the right (R) and left (L) valleys. The effective Hamiltonian describing the Dirac electron system in α-(BEDT-TTF)$_2$I$_3$ is given by [10,31]

$$H = H_0 + H', \tag{1}$$

where kinetic energy term H_0 is

$$H_0 = \sum_{q\gamma\gamma'\tau s} a^\dagger_{\mathbf{r}\gamma,\tau s}[\hat{H}_0^{\tau s}]_{\gamma\gamma'} a_{\mathbf{r}\gamma',\tau s}, \tag{2}$$

$$\hat{H}_0^{\tau s} = -i\tau\hbar v \begin{pmatrix} \eta\partial_x & \partial_x - i\tau\partial_y \\ \partial_x + i\tau\partial_y & \eta\partial_x \end{pmatrix}, \tag{3}$$

where $a^\dagger_{\mathbf{r}\gamma,\tau s}$ and $a_{\mathbf{r}\gamma,\tau s}$ represent creation and annihilation operators, respectively, with two dimensional space $\mathbf{r} = (x,y)$, Luttinger–Kohn base γ [10,37], valley $\tau = \pm$ (R, L), and spin $s = \pm$ ($\uparrow$, $\downarrow$). The degree of tilt η is defined by $\eta = v_0/v$ with cone velocity v and tilt velocity v_0, where the anisotropy of the cone velocity is ignored here for simplicity [31]. The energy eigenvalue of H_0 is given by

$$E = \hbar v \left(\eta k_x \pm \sqrt{k_x^2 + k_y^2} \right), \tag{4}$$

with two-dimensional momentum k_x and k_y. The left inset in Figure 1 shows the tilted Dirac cone, where $v \pm v_0$ are velocities in the $\pm k_x$ directions, respectively. Interaction term $\hat{H}'$ is given by

$$\hat{H}' = \frac{1}{2} \int d\mathbf{r} \int d\mathbf{r}' V(\mathbf{r} - \mathbf{r}') n(\mathbf{r}) n(\mathbf{r}') \tag{5}$$

with long-range Coulomb interaction $V(\mathbf{r}) = e^2/\varepsilon r$ and electron density operator $n(\mathbf{r})$.

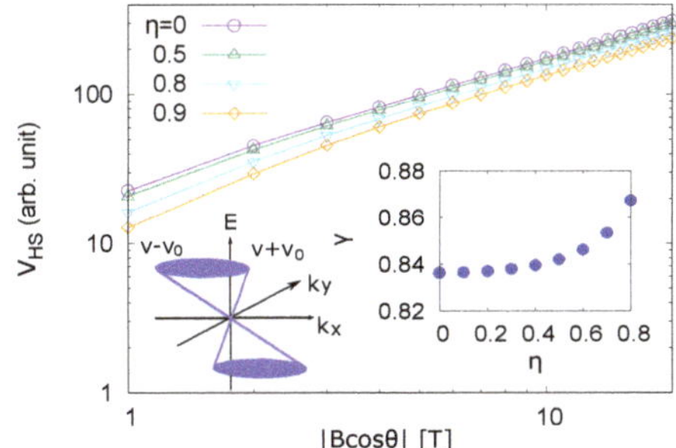

Figure 1. Theoretical results on $|B\cos\theta|$ dependences of V_{HS} for $\eta = 0$ (circle), 0.5 (triangle), 0.8 (inverted triangle), and 0.9 (diamond). The right inset shows the η dependence of γ. The left inset indicates the tilted Dirac cone, where v and v_0 are cone velocity and tilt velocity, respectively.

The energy spectrum of the two-dimensional electronic system is quantized under a tilted magnetic field, $\mathbf{B} = \nabla \times \mathbf{A} = (B_x, B_y, B_z)$, given by the vector potential with the Landau gauge, $\mathbf{A} = (B_y z - B_z y, -B_x z, 0)$. The Hamiltonian under the magnetic field is obtained by Peierls substitution $-i\nabla \to -i\nabla + (e/\hbar)\mathbf{A}$ and Zeeman energy $E_Z = g\mu_B B/2$ with $B = |\mathbf{B}|$. The Landau levels are given by $E_{Ns} = \mathrm{sgn}(N)(\hbar v/l)\sqrt{2\lambda^3|N|} - sE_Z$ with $\lambda = \sqrt{1-\eta^2}$ and magnetic length $l = \sqrt{\hbar/e|B_z|}$. The wave function $\mathbf{\Psi}^\tau_{N,k}(\mathbf{r})$ for the $N = 0$ Landau level is given by References [11,27,38]

$$\mathbf{\Psi}^{\tau=+}_{N=0,k}(\mathbf{r}) = \frac{1}{\sqrt{2(1+\lambda)}} \begin{pmatrix} -\eta \\ 1+\lambda \end{pmatrix} \phi_{N=0,k}(\mathbf{r}), \tag{6}$$

$$\mathbf{\Psi}^{\tau=-}_{N=0,k}(\mathbf{r}) = \frac{1}{\sqrt{2(1+\lambda)}} \begin{pmatrix} -1-\lambda \\ \eta \end{pmatrix} \phi_{N=0,k}(\mathbf{r}), \tag{7}$$

where

$$\phi_{N=0,k}(\mathbf{r}) = \frac{\lambda^{1/4}}{\sqrt{2\pi}} \frac{1}{(\sqrt{\pi}l)^{1/2}} e^{ikx} e^{-\tilde{Y}^2/2} \tag{8}$$

with $\tilde{Y} = \frac{\sqrt{\lambda}}{l}(y - Y)$ and center coordinate $Y = l^2 k$. When $H = 5$ T, $E_1 \cong 3.5$ meV using values of parameters for α-(BEDT-TTF)$_2$I$_3$, where velocity $v = 1.0 \times 10^5$ m/s and $v_0 = 0.8 \times 10^5$ m/s are given by the band calculation [10,23,31], $2E_Z \cong 0.5$ meV with $g = 2$, $l \cong 14$ nm.

In the present study, we consider the $N = 0$ Landau levels in order to study interlayer magnetoresistivity in the quantum limit and, for the case of $E_1 \gg E_Z$. The effective g-factor, g^*, is treated by the mean field theory. The Green function $G_s(k, i\varepsilon_n)$ is defined by

$$G_s(k, i\varepsilon_n) = \frac{1}{i\varepsilon_n + \mu + sE_Z - \Sigma_s}, \tag{9}$$

where $\epsilon_n = (2n+1)\pi/\beta$ is fermion Matsubara frequency with $\beta = 1/k_B T$ and integer n, and μ is the chemical potential, determined so that the Fermi energy coincides with the Dirac point. The self-energy Σ_s is given by self-consistent equation

$$\Sigma_\uparrow - \Sigma_\downarrow = -V_{HS} m_z, \tag{10}$$

with effective Coulomb interaction $V_{HS} = \sum_{\tau'}[V(0)]_{\tau,\tau,\tau',\tau'}$, magnetization $m_z = n_\uparrow - n_\downarrow$, and electron density for spin s, n_s, where

$$\Sigma_s = \frac{1}{D}\sum_{k'}\frac{1}{\beta}\sum_{\varepsilon'_n}\sum_{\tau'}[V(0)]_{\tau,\tau,\tau',\tau'} G_{-s}(k', i\varepsilon'_n) \tag{11}$$

and the Coulomb interaction matrix $[V(q)]_{\tau_1\tau_2,\tau_3,\tau_4}$ for the $N=0$ Landau levels is given by

$$[V(q)]_{\tau_1\tau_2,\tau_3,\tau_4} = \frac{1}{2}\int d\mathbf{r}\int d\mathbf{r}'V(\mathbf{r}-\mathbf{r}')\times$$
$$[\mathbf{\Psi}_{N=0,k+q}^{\tau_2}(\mathbf{r})^{\dagger}\cdot\mathbf{\Psi}_{N=0,k}^{\tau_1}(\mathbf{r})][\mathbf{\Psi}_{N=0,k'-q}^{\tau_4}(\mathbf{r}')^{\dagger}\cdot\mathbf{\Psi}_{N=0,k'}^{\tau_3}(\mathbf{r}')], \tag{12}$$

where the degeneracy of center coordinate $D = L_x L_y / 2\pi l^2$ with length L_x and L_y in the x and y directions, respectively. The effective Coulomb interaction depends on magnetic length l, so effective Coulomb interaction depends on angle θ. Effective spin splitting E_Z^* is given by

$$E_Z^* = g^*\mu_B B/2 = E_Z + \Delta \tag{13}$$

with $\Delta = \Sigma_\downarrow = -\Sigma_\uparrow \geq 0$ for the charge neutral system $\mu = 0$. Thus, effective g-factor g^* is given by

$$g^* = g + 2\Delta/\mu_B B. \tag{14}$$

Thus, the energy eigenvalues of $N = 0$ Landau levels are modified as follows:

$$E_{N=0,s} = -sE_Z^* = -s(E_Z + \Delta). \tag{15}$$

Interlayer conductivity is given by interlayer coupling as a perturbation [25]. The perturbation Hamiltonian H' is given by

$$\hat{H}' = -2t_c \cos\left(-ic\frac{\partial}{\partial z}\right)\begin{pmatrix}1 & 0 \\ 0 & 1\end{pmatrix} \tag{16}$$

where t_c and c represent interlayer transfer energy and interlayer spacing, respectively. In the quantum limit, $N = 0$ Landau levels are dominant in magnetotransport. The effective transfer energy between $N = 0$ Landau levels in neighboring layers is given by [25]:

$$\tilde{t}_c(Y',z_i';Y,z_i) = t_c \times \exp\left[-\frac{1}{4}\frac{c^2 e\left(B_x^2 + B_y^2\right)}{\hbar|B_z|}\right]$$
$$\times \exp\left[i\frac{eB_x}{\hbar}(z_i'-z_i)\frac{Y+Y'}{2}\right], \tag{17}$$

where z_i is the layer position, and the neighboring layers are $z_i' = z_i \pm c$. The center coordinate of initial state Y on one layer $z = z_i$, and that of final state Y' on neighboring layer $z = z_i \pm c$ satisfy the condition

$$Y' = Y \pm \frac{B_y}{B_z}c. \tag{18}$$

The complex interlayer conductivity $\tilde{\sigma}_{zz}(\omega)$ is given by [25]:

$$\tilde{\sigma}_{zz}(\omega) = -\frac{i\hbar}{L^2}\sum_{Y,z_i,\tau,s;Y',z_i',\tau',s'}\langle N=0,Y,z_i,\tau,s\,|\,\hat{j}_z\,|\,N=0,Y',z_i',\tau',s'\rangle$$
$$\times\langle N=0,Y',z_i',\tau',s'\,|\,\hat{j}_z\,|\,N=0,Y,z_i,\tau,s\rangle$$
$$\times\frac{f(E_{N=0,s})-f(E_{N=0,s'})}{E_{N=0,s'}-E_{N=0,s}}\frac{1}{E_{N=0,s'}-E_{N=0,s}-\hbar\omega-i\hbar/\tilde{\tau}'}, \tag{19}$$

where $\tilde{\tau}$ and $\hat{j}_z$ represent relaxation time and interlayer current density, respectively. The center coordinate Y is associated with wavenumber k by relation expression $Y = l^2 k$. Interlayer current density is given by $\hat{j}_z = (-e)(1/i\hbar)[\hat{z}, \hat{H}']$.

The leading term of real conductivity is given by [25]:

$$\sigma_{zz} = \frac{2Ct_c^2 ce^3 \tilde{\tau} |B_z|}{\pi \hbar^3} \exp\left[-\frac{ec^2(B_x^2 + B_y^2)}{2\hbar |B_z|} \right] \tag{20}$$

with

$$C = \sum_{s\tau} \int D_{s\tau}(E)^2 \left(-\frac{\mathrm{d}f}{\mathrm{d}E} \right) \mathrm{d}E \tag{21}$$

where $D_{s\tau}(E)$, t_c and c denote density of states for spin s and valley τ, interlayer transfer energy, and interlayer spacing, respectively. Interlayer resistivity ρ_{zz} is given by $\rho_{zz} \simeq 1/\sigma_{zz}$, since interlayer Hall conductivities σ_{xz} and σ_{yz} are negligibly smaller than other components [39,40].

2.2. Experimental Method

2.2.1. Crystal Growth

Either the electrolysis method or the diffusion method are generally used in the crystal growth of organic conductors. We synthesized organic conductor α-(BEDT-TTF)$_2$I$_3$ by the electrolysis method using an H-type cell. BEDT-TTF molecules and I$_3^-$ anions are desolved in a supporting electrolyte (THF, benzonitrile, chlorobenzene, etc.). Then, we supply electrical current (1$\sim$5 μA) between platinum electrodes. After about 1$\sim$2 weeks, small single crystals appear on the positively based platinum electrode. The typical size of a crystal is $1 \times 0.5 \times 0.05$ mm^3.

2.2.2. Experiments of Interlayer Magnetoresistance Under Pressure

A sample with a dimension of approximately $0.7 \times 0.5 \times 0.05$ mm^3, on which four electrical leads (gold wire with a diameter of 15 μm) are attached by carbon paste, is put in a Teflon capsule filled with the pressure medium (Idemitsu DN-oil 7373); then, the capsule is set in a NiCrAl clamp cell, and hydrostatic pressure of up to 1.7 GPa is applied. The hydrostatic pressure is determined at room temperature by a Manganine resistance gauge in the pressure cell. The interlayer resistance of a crystal is measured by a conventional DC method with an electrical current of 1 μA along the c-crystal axis, which is normal to the two-dimensional plane. In the investigation, interlayer magnetoresistance is measured as functions of B and θ which is the angle between the magnetic-field direction and c-crystal axis at $T = 0.5, 1.7, 2.5, 3.0, 3.5$, and 4.2 K. As mentioned in the introduction, interlayer magnetoresistance is a useful tool to detect the effects of $N = 0$ Landau levels, including its Zeeman splittings.

3. Results

Figure 1 shows the theoretical results on $|B\cos\theta|$ $(= |B_z|)$ dependences of effective Coulomb interaction V_{HS} for tilt parameter $\eta = 0, 0.5, 0.8$, and 0.9. It is numerically shown that $V_{HS} \propto |B\cos\theta|^\gamma$ approximately for $|B\cos\theta| > 3$. Effective Coulomb interaction depends on magnetic length l, which is a function of $|B_z|$, as shown in Equation (12). Thus, effective Coulomb interaction depends on angle θ. The left inset shows the tilting Dirac cone, where v and v_0 represent cone velocity and tilt velocity, respectively. The right inset shows the η dependence of γ, where γ increases as η increases. Thus, we use a relation, $V_{HS} = u|B\cos\theta|^\gamma$, with $\gamma = 0.87$ for the tilted Dirac cones in α-(BEDT-TTF)$_2$I$_3$ with $\eta = 0.8$ [23] hereafter. Parameter $u = 0.3$ is chosen to fit with the experimental results.

Figure 2a,b shows the theoretical results on the B-dependences of g^* and E_Z^*, respectively, for $\theta = 0$, 20, 40, and $60°$ at $T = 1.7$ K. It is found that both g^* and E_Z^* enhance by V_{HS} depend on θ. When

$\theta = 90°$, $g^* = g$ and $E_Z^* = E_Z$, since $V_{HS} = 0$. Although E_Z^* increases monotonically as B increases, g^* has a maximum since Δ divided by B contributes to g^*.

Figure 3a shows the theoretical results on B-dependences of interlayer resistivity ρ_{zz} for $\theta = 0$, 20, 40, and 60° at $T = 1.7$ K. ρ_{zz} has a minimum at B_0. It is found that B_0 increases as θ increases due to θ-dependence of E_Z^*. Figure 3b shows the experimental results on B-dependences of interlayer resistance R_{zz} for $\theta = 0$, 20, 40, and 60° at $T = 1.7$ K. B_0 obtained in the experimental results also increases as θ increases.

In a general two-dimensional system, E_Z^* dose not depend on θ. The agreement between the theory and experiment of interlayer magnetoresistance shown in Figure 3 indicates that effective Coulomb interaction plays an important role to the Zeeman effects. In the following, peculiar Zeeman effects in this system are examined.

The first step is to investigate g^* for $\theta = 0$ from the interlayer-magnetoresistance minimum, where $g^* \mu_B B_0 \sim 2\hbar/\tilde{\tau}$. Here, the rough of broadening energy $\hbar/\tilde{\tau}$ of the Landau levels in this system is 3 K at low temperatures [32]. g^* for $\theta = 0$ is roughly estimated to be 5 experimentally at 1.7 K; this is close to the theoretical value, which is approximately 5.3 at $B_0 \sim 1.8$ T, as shown in Figure 2a.

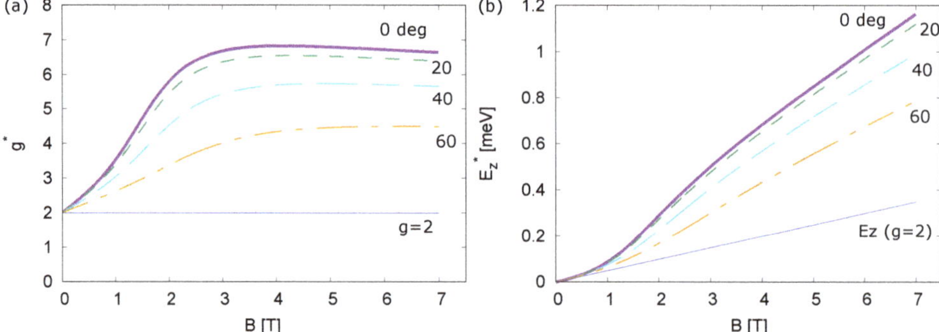

Figure 2. Numerical results on B-dependences of g^* (**a**) and E_Z^* (**b**), respectively, at $T = 1.7$ K for $\theta = 0$ (solid line), 20 (dotted line), 40 (dashed line), and 60° (dot-dashed line). g and E_Z in the absence of interaction are drawn by the thin solid line.

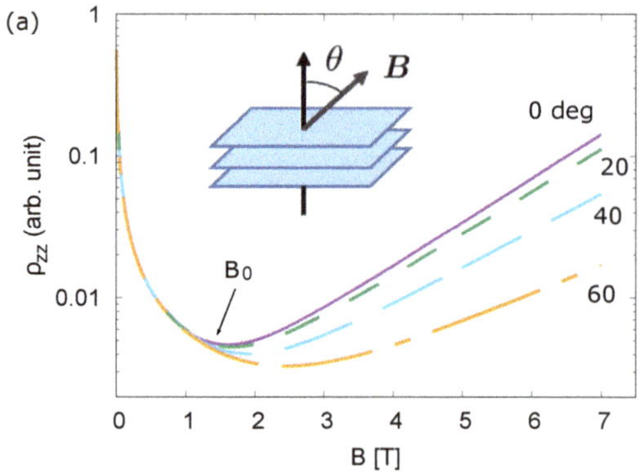

Figure 3. *Cont.*

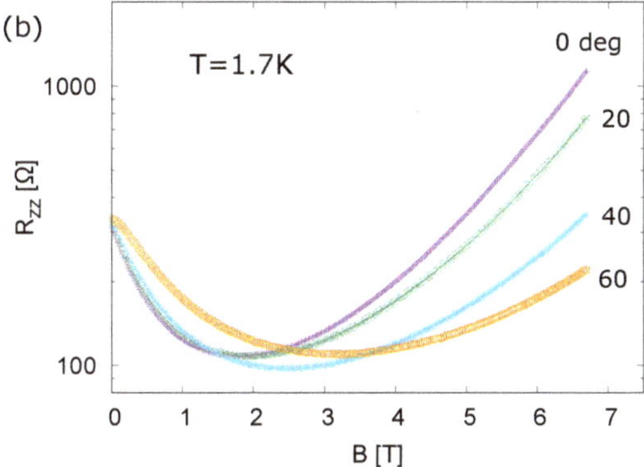

Figure 3. (**a**) theoretical results on B-dependences of ρ_{zz} for $\theta = 0$ (solid line), 20 (dotted line), 40 (dashed line), and 60° (dot-dashed line) at $T = 1.7$ K. (**b**) experimental results on B-dependences of R_{zz} for $\theta = 0, 20, 40,$ and 60° at $T = 1.7$ K.

Lastly, the peculiar Zeeman effects on the θ-dependence of the interlayer magnetoresistance minimum are detected. In Figure 4a, the theoretical results on θ dependences of B_0^{-1} are drawn as functions of $\cos\theta$ for $T = 0.5, 1.7, 2.5, 3.0, 3.5,$ and 4.2 K. It is numerically shown that $1/B_0 \cong a\cos\theta + b$ when θ is close to 0°. Note that this numerical calculation is not applicable when θ is close to 90°, where contributions of higher Landau levels in interlayer resistivity are not negligible. Figure 4b shows the experimental results on $\cos\theta$ dependences of B_0^{-1} for $T = 0.5, 1.7, 2.5, 3.0, 3.5,$ and 4.2 K. The experimental results also show the same approximate relation on B_0^{-1} as a linear function of $\cos\theta$.

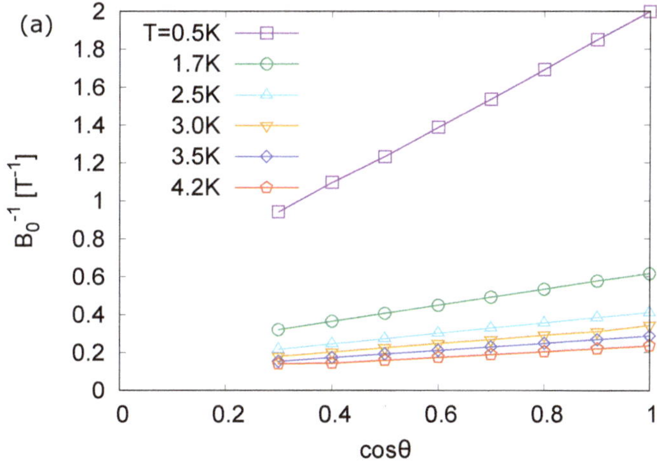

Figure 4. *Cont.*

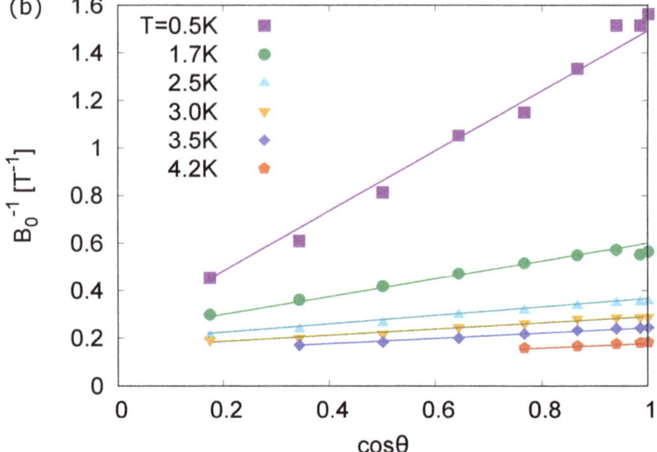

Figure 4. (**a**) theoretical results on $\cos\theta$ dependences of B_0^{-1} for $T = 0.5$ (open square), 1.7 (open circle), 2.5 (open triangle), 3.0 (inverted open triangle), 3.5 (open diamond), and 4.2 K (open pentagon). When $\cos\theta$ is close to 1, numerical results show a relation $1/B_0 \cong a\cos\theta + b$ approximately, where a and b are independent of θ. (**b**) experimental results on $\cos\theta$ dependences of B_0^{-1} for $T = 0.5$ (filled square), 1.7 (filled circle), 2.5 (filled triangle), 3.0 (inverted filled triangle), 3.5 (filled diamond), and 4.2 K (filled pentagon). As with the theoretical results, the experimental results show relation $1/B_0 \cong a\cos\theta + b$ approximately when $\cos\theta$ is close to 1.

Figure 5 shows the theoretical and experimental results on T dependencies of coefficient a. It is found that coefficient a increases as T decreases, indicating remarkable θ dependence of B_0^{-1} at very low temperatures $T \leq 1.7$ K. Those results show qualitative agreement between theory and experimental results for the field-angle dependence of interlayer magnetoresistance.

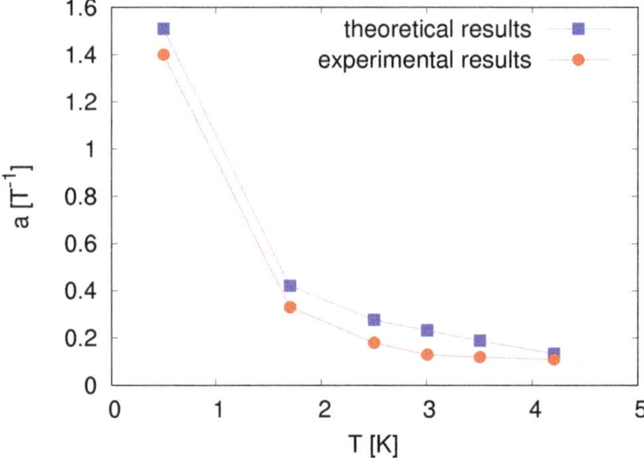

Figure 5. Theoretical results (filled square) and experimental results (filled circle) of coefficient a as a function of T in relation $1/B_0 \cong a\cos\theta + b$. Coefficient a increases as T decreases. Coefficient a especially rapidly increases at low temperatures. Theoretical and experimental results coincide.

4. Discussion

Those theoretical curves in Figures 3a and 4a qualitatively reproduce the experimental results including the characteristic features of the resistance minimum as shown in Figures 3b and 4b. Here, the theoretical results do not coincide with the experimental results near $B = 0$ T, since only $N = 0$ Landau levels are examined in the present theoretical calculation. This theoretical approach, however, is valid because the special Landau-level structure in this system realizes the quantum limit in the magnetic field above 0.07 T for a perpendicular component to the two-dimensional plane at 1.7 K [32,41].

In the present study, the value of u for V_{HS} is chosen to fit with the experimental results. The value of u will be evaluated by taking the Thomas–Fermi screening effect into account. The effects of the Coulomb interaction in $N \neq 0$ Landau levels is also going to be studied in the mean field calculation. In addition, interlayer magnetoresistance is also going to be studied, taking $N \neq 0$ Landau levels into account, which leads to the maximum of the interlayer resistance under weak magnetic fields [26,28].

5. Conclusions

In conclusion, the effect of the Coulomb interaction in interlayer magnetoresistance was elucidated in collaboration with theory and experiments for the Dirac electron system in organic conductor α-(BEDT-TTF)$_2$I$_3$. It was shown that enhancement of the effective g-factor due to the Coulomb interaction for $N = 0$ Landau levels depends on field angle θ. This was the key factor to explain the field-angle dependence of B_0 at which interlayer resistance has a minimum.

The value of the effective g-factor obtained in the present study at $\theta = 0$ under a strong magnetic field is consistent with the theoretical and experimental results of the Nernst effect [42,43]. This is also consistent with the theoretical and experimental results of the spin-lattice relaxation rate [33,36]. The quantum Hall ferromagnetic state assumed in the present study supports the helical surface state [29].

Author Contributions: T.T. and A.K. conceived and designed the theories; T.T. and A.K. performed the numerical calculations and analyzed the data; N.T. performed the experiments and analyzed the data; T.T., N.T., and A.K. wrote the paper.

Funding: This work was supported by MEXT/JSPJ KAKENHI under Grant Nos. 15K05166, 18K18739, and 16H06346.

Acknowledgments: The authors would like to thank S. Hayashi for the fruitful discussions.

Conflicts of Interest: The authors declare no conflict of interest.

Abbreviations

The following abbreviations are used in this manuscript:

R   right
L   left

References

1. Goerbig, M.O. Electronic properties of graphene in a strong magnetic field. *Rev. Mod. Phys.* **2011**, *83*, 1193. [CrossRef]
2. Suzumura, Y.; Kobayashi, A. Theory of Dirac Electrons in Organic Conductors. *Crystals* **2012**, *2*, 266. [CrossRef]
3. Kotov, V.N.; Uchoa, B.; Pereira, V.M.; Guinea, F.; Castro Neto, A.H. Electron-Electron Interactions in Graphene: Current Status and Perspectives. *Rev. Mod. Phys.* **2012**, *84*, 1067. [CrossRef]
4. Wehling, T.O.; Black-Schaffer, A.M.; Balatsky, A.V. Dirac materials. *Adv. Phys.* **2014**, *63*, 1. [CrossRef]
5. Kajita, K.; Nishio, Y.; Tajima, N.; Suzumura, Y.; Kobayashi, A. Molecular Dirac Fermion Systems -Theoretical and Experimental Approaches-. *J. Phys. Soc. Jpn.* **2014**, *83*, 072002. [CrossRef]

6. Kajita, K.; Ojiro, T.; Fujii, H.; Nishio, Y.; Kobayashi, H.; Kobayashi, A.; Kato, R. Magnetotransport Phenomena of α-Type (BEDT-TTF)$_2$I$_3$ under High Pressures. *J. Phys. Soc. Jpn.* **1992**, *61*, 23. [CrossRef]
7. Tajima, N.; Tamura, M.; Nishio, Y.; Kajita, K.; Iye, Y. Transport Property of an Organic Conductor α-(BEDT-TTF)$_2$I$_3$ under High Pressure -Discovery of a Novel Type of Conductor-. *J. Phys. Soc. Jpn.* **2000**, *69*, 543. [CrossRef]
8. Kobayashi, A.; Katayama, S.; Noguchi, K.; Suzumura, Y. Superconductivity in Charge Ordered Organic Conductor α-(ET)$_2$I$_3$ Salt. *J. Phys. Soc. Jpn.* **2004**, *73*, 3135. [CrossRef]
9. Katayama, S.; Kobayashi, A.; Suzumura, Y. Pressure-Induced Zero-Gap Semiconducting State in Organic Conductor α-(BEDT-TTF)$_2$I$_3$ Salt. *J. Phys. Soc. Jpn.* **2006**, *75*, 054705. [CrossRef]
10. Kobayashi, A.; Katayama, S.; Suzumura, Y.; Fukuyama, H. Massless Fermions in Organic Conductor. *J. Phys. Soc. Jpn.* **2007**, *76*, 034711. [CrossRef]
11. Goerbig, M.O.; Fuchs, J.-N.; Montambaux, G.; Piechon, F. Tilted anisotropic Dirac cones in quinoid-type graphene and α-(BEDT-TTF)$_2$I$_3$. *Phys. Rev. B* **2008**, *78*, 045415. [CrossRef]
12. Kobayashi, A.; Katayama, S.; Suzumura, Y. Theoretical study of the zero-gap organic conductor α-(BEDT-TTF)$_2$I$_3$. *Sci. Technol. Adv. Mater.* **2009**, *10*, 024309. [CrossRef]
13. Tajima, N.; Kajita, K. Experimental study of organic zero-gap conductor α-(BEDT-TTF)$_2$I$_3$. *Sci. Technol. Adv. Mater.* **2009**, *10*, 024308. [CrossRef]
14. Kino, H.; Fukuyama, H. Interrelationship among Electronic States of α-(ET)$_2$I$_3$, (ET)$_2$MHg(SCN)$_4$ and κ-(ET)$_2$X. *J. Phys. Soc. Jpn.* **1995**, *64*, 4523. [CrossRef]
15. Seo, H. Charge Ordering in Organic ET Compounds. *J. Phys. Soc. Jpn.* **2000**, *69*, 805. [CrossRef]
16. Kobayashi, A.; Suzumura, Y.; Piechon, F.; Montambaux, G. Emergence of Dirac electron pair in the charge-ordered state of the organic conductor α-(BEDT-TTF)$_2$I$_3$. *Phys. Rev. B* **2011**, *84*, 075450. [CrossRef]
17. Tanaka, Y.; Ogata, M. Correlation Effects on Charge Order and Zero-Gap State in the Organic Conductor α-(BEDT-TTF)$_2$I$_3$. *J. Phys. Soc. Jpn.* **2016**, *85*, 104706. [CrossRef]
18. Hirata, M.; Ishikawa, K.; Miyagawa, K.; Tamura, M.; Berthier, C.; Basko, D.; Kobayashi, A.; Matsuno, G.; Kanoda, K. Observation of an anisotropic Dirac cone reshaping and ferrimagnetic spin polarization in an organic conductor. *Nat. Commun.* **2016**, *7*, 12666. [CrossRef]
19. Matsuno, G.; Kobayashi, A. Effect of Interband Fluctuation on Spin Susceptibility in Molecular Dirac Fermion System α-(BEDT-TTF)$_2$I$_3$. *J. Phys. Soc. Jpn.* **2017**, *86*, 014705. [CrossRef]
20. Hirata, M.; Ishikawa, K.; Matsuno, G.; Kobayashi, A.; Miyagawa, K.; Tamura, M.; Berthier, C.; Kanoda, K. Anomalous spin correlations and excitonic instability of interacting 2D Weyl fermions. *Science* **2017**, *358*, 1403. [CrossRef] [PubMed]
21. Matsuno, G.; Kobayashi, A. Coexistence of Velocity Renormalization and Ferrimagnetic Fluctuation in the Organic Dirac Electron System α-(BEDT-TTF)$_2$I$_3$. *J. Phys. Soc. Jpn.* **2018**, *87*, 054706. [CrossRef]
22. Ohki, D.; Matsuno, G.; Omori, Y.; Kobayashi, A. Optical Conductivity in a Two-Dimensional Extended Hubbard Model for an Organic Dirac Electron System α-(BEDT-TTF)$_2$I$_3$. *Crystals* **2018**, *8*, 137. [CrossRef]
23. Kobayashi, A.; Suzumura, Y.; Fukuyama, H. Hall Effect and Orbital Diamagnetism in Zerogap State of Molecular Conductor α-(BEDT-TTF)$_2$I$_3$. *J. Phys. Soc. Jpn.* **2008**, *77*, 064718. [CrossRef]
24. Tajima, N.; Kato, R.; Sugawara, S.; Nishio, Y.; Kajita, K. Interband effects of magnetic field on Hall conductivity in the multilayered massless Dirac fermion system α-(BEDT-TTF)$_2$I$_3$. *Phys. Rev. B* **2012**, *85*, 033401. [CrossRef]
25. Osada, T. Negative Interlayer Magnetoresistance and Zero-Mode Landau Level in Multilayer Dirac Electron Systems. *J. Phys. Soc. Jpn.* **2008**, *77*, 084711. [CrossRef]
26. Tajima, N.; Sugawara, S.; Kato, R.; Nishio, Y.; Kajita, K. Effect of the Zero-Mode Landau Level on Interlayer Magnetoresistance in Multilayer Massless Dirac Fermion Systems. *Phys. Rev. Lett.* **2013**, *102*, 176403, [CrossRef] [PubMed]
27. Morinari, T.; Himura, T.; Tohyama, T. Possible Verification of Tilted Anisotropic Dirac Cone in α-(BEDT-TTF)$_2$I$_3$ Using Interlayer Magnetoresistance. *J. Phys. Soc. Jpn.* **2009**, *78*, 023704. [CrossRef]
28. Morinari, T.; Tohyama, T. Crossover from Positive to Negative Interlayer Magnetoresistance in Multilayer Massless Dirac Fermion System with Non-Vertical Interlayer Tunneling. *J. Phys. Soc. Jpn.* **2010**, *79*, 044708. [CrossRef]
29. Osada, T. Surface Transport in the $\nu = 0$ Quantum Hall Ferromagnetic State in the Organic Dirac Fermion System. *J. Phys. Soc. Jpn.* **2015**, *84*, 053704. [CrossRef]

30. Tajima, N.; Morinari, T. Tilted Dirac Cone Effect on Interlayer Magnetoresistance in α-(BEDT-TTF)$_2$I$_3$. *J. Phys. Soc. Jpn.* **2018**, *87*, 045002. [CrossRef]
31. Kobayashi, A.; Suzumura, Y.; Fukuyama, H.; Goerbig, M.O. Tilted-Cone-Induced Easy-Plane Pseudo-Spin Ferromagnet and Kosterlitz Thouless Transition in Massless Dirac Fermions. *J. Phys. Soc. Jpn.* **2009**, *78*, 114711. [CrossRef]
32. Tajima, N.; Sato, M.; Sugawara, S.; Kato, R.; Nishio, Y.; Kajita, K. Spin and valley splittings in multilayered massless Dirac fermion system. *Phys. Rev. B* **2010**, *82*, 121420(R). [CrossRef]
33. Hirata, M. NMR Studies of Massless Dirac Fermions in the Quasi-Two-Dimensional Organic Conductor α-(BEDT-TTF)$_2$I$_3$. Ph.D. Thesis, The University of Tokyo, Tokyo, Japan, 2012.
34. Kubo, K.; Morinari, T. Spin-Ordered States in Multilayer Massless Dirac Fermion Systems. *J. Phys. Soc. Jpn.* **2014**, *83*, 033702. [CrossRef]
35. Kubo, K.; Morinari, T. Effect of Interlayer Spin-Flip Tunneling for Interlayer Magnetoresistance in Multilayer Massless Dirac Fermion Systems. *J. Phys. Soc. Jpn.* **2014**, *83*, 083701. [CrossRef]
36. Tani, T.; Kobayashi, A. Spin-Lattice Relaxation Rate in Organic Dirac Electron System α-(BEDT-TTF)$_2$I$_3$ under Strong Magnetic Field. *J. Phys. Soc. Jpn.* **2019**, *88*, in press.
37. Luttinger, J.M.; Kohn, W. Motion of Electrons and Holes in Perturbed Periodic Fields. *Phys. Rev.* **1955**, *97*, 869. [CrossRef]
38. Sári, J.; Töke, C.; Goerbig, M.O. Magnetoplasmons of the tilted anisotropic Dirac cone material α-(BEDT-TTF)$_2$I$_3$. *Phys. Rev. B* **2014**, *90*, 155446. [CrossRef]
39. Osada, T. Anomalous Interlayer Hall Effect in Multilayer Massless Dirac Fermion System at the Quantum Limit. *J. Phys. Soc. Jpn.* **2011**, *80*, 033708. [CrossRef]
40. Osada, T. Magnetotransport in organic Diracfermion system at the quantum limit:Interlayer Hall effect and surfacetransport via helical edge states. *Phys. Status Solidi* **2012**, *249*, 962–966. [CrossRef]
41. Sugawara, S.; Tamura, M.; Tajima, N.; Kato, R.; Sato, M.; Nishio, Y.; Kajita, K. Temperature Dependence of Inter-Layer Longitudinal Magnetoresistance in α-(BEDT-TTF)$_2$I$_3$: Positive versus Negative Contributions in a Tilted Dirac Cone System. *J. Phys. Soc. Jpn.* **2010**, *79*, 113704. [CrossRef]
42. Proskurin I.; Ogata, M. Thermoelectric Transport Coefficients for Massless Dirac Electrons in Quantum Limit. *J. Phys. Soc. Jpn.* **2013**, *82*, 063712. [CrossRef]
43. Konoike, T.; Sato, M.; Uchida, K.; Osada, T. Anomalous Thermoelectric Transport and Giant Nernst Effect in Multilayered Massless Dirac Fermion System. *J. Phys. Soc. Jpn.* **2013**, *82*, 073601. [CrossRef]

© 2019 by the authors. Licensee MDPI, Basel, Switzerland. This article is an open access article distributed under the terms and conditions of the Creative Commons Attribution (CC BY) license (http://creativecommons.org/licenses/by/4.0/).

Article

Synthesis of ErBa$_2$Cu$_3$O$_{7-\delta}$ Superconductor Solder for the Fabrication of Superconducting Joint between Gdba$_2$cu$_3$o$_{7-\delta}$ Coated Conductor

Zili Zhang [1], Lei Wang [1], Jianhua Liu [1] and Qiuliang Wang [1,2,*

[1]   Institute of Electrical Engineering, Chinese Academy of Sciences, Beijing 100190, China; zilizhang0816@vip.126.com (Z.Z.); wanglei@mail.iee.ac.cn (L.W.); liujianhua@mail.iee.ac.cn (J.L.)

[2]   University of Chinese Academy of Sciences, Beijing 100049, China

*   Correspondence: qiuliang@mail.iee.ac.cn

Received: 3 September 2019; Accepted: 21 September 2019; Published: 25 September 2019

Abstract: ErBa$_2$Cu$_3$O$_{7-\delta}$ (Er123) superconductor is one of the best candidates of superconductor solder for the fabrication of superconducting joint between GdBa$_2$Cu$_3$O$_{7-\delta}$ (Gd123) coated conductor, due to its high T_c value (93 K) and highest optimized oxygen annealing temperature among RE123 compounds. In this paper, we systematically research the effect of sintering parameters on the phase formation, microstructure and superconducting properties of Er123 powder. The optimized synthesis route to acquire high purity Er123 powder with as good superconducting properties as Gd123 has been uncovered. The melt temperature of Er123 with different dopant compared to Gd123 is also investigated, and the feasible operating temperature range of Er123 superconductor solder is discussed. This work reveals a very important starting point on fabrication high-quality superconducting joint between the commercial Gd123 coated conductor, which can further improve the development of the persistent operating mode on ultra-high field nuclear magnetic resonance and magnetic resonance imaging.

Keywords: Er123; melt temperature; superconducting solder; superconducting joint

---

1. Introduction

REBa$_2$Cu$_3$O$_{7-\delta}$ (REBCO) has been considered to be one of the promising superconductors for the insert coil of ultra-high field nuclear magnetic resonance and magnetic resonance imaging [1–5]. The RIKEN has successfully fabricated an NMR by REBCO operated in the driven mode and demonstrated high-resolution NMR spectra [6]. In the real application, the persistent mode is preferred to reduce the heat leak and get a more stable magnetic field. One of the key points of persistence mode is a superconducting joint between REBCO coated conductor with resistance less than 10^{-12} Ω [7]. However, most joints now are on the level 10^{-8} to 10^{-9} Ω [8–10], which indicates that a feasible superconducting joint fabrication process is still required.

The first persistent current joint between REBCO was invented by Park et al. in 2014 [11]. The GdBa$_2$Cu$_3$O$_{7-\delta}$ (Gd123) coated conductor was directly connected by long time heat treatment to diffuse the Gd123 to each other. After over 350 h oxygen annealing, the final joint had a critical current of 84 A and a resistance less than 10^{-17} Ω. Although the properties fit the application criterion of the persistent mode, the too long annealing time is not feasible in the real application. In 2015, the Jin et al. in RIKEN established a novel method called crystalline joint by a melted bulk (CJMB). Two untouched Gd123 coated conductors were put on a RE123 bulk with low melt temperatures such as YBa$_2$Cu$_3$O$_{7-\delta}$ (Y123) and YbBa$_2$Cu$_3$O$_{7-\delta}$ (Yb123) [12,13]. After ingeniously designed the heat treatment, the Y123 melt and regrowth to form a superconductor joint between Gd123 coated conductors. The annealing time was successfully reduced to only 72 h with a final resistance of 8×10^{-13} Ω. The core idea of this method is using low melt temperature REBCO materials as the superconducting solder.

Using low melt temperature REBCO as a superconducting solder has also been used in joining Y123 single domain bulks [14–18]. In all these researches, appropriate oxygen annealing is essential to achieve high supercurrent capability [19]. In particular, $ErBa_2Cu_3O_{7-\delta}$ (Er123) has the highest optimized annealing temperature among RE123 compounds, which is strongly suggesting a drastic decrease in annealing time for the accomplishment of an optimally carrier-doped condition in Er123 compounds due to the large diffusion coefficient of oxygen [20]. There are only a few reports mentioned on Er123 powder [21–26], and none of them has systematic research on the synthesis of such compounds. The reported synthesis temperature even has some tens of degrees difference, which brings big confusion to other researchers.

In this paper, we systematically investigate the effect of synthesis parameters on the phase formation, microstructure and superconducting properties. High purity Er123 powder with high supercurrent capability is acquired. The melt temperature of Er123 and Gd123 with different dopants are also researched.

2. Materials and Methods

Powder Preparation. The Er123 powders were synthesized by the solid reaction method. The Er_2O_3 (99.9% Aladdin), $BaCO_3$ (99.9% Aladdin) and CuO powder (99.99% Aladdin) powder were mixed in stoichiometric ratio to form Er123. The powders were mixed, ground together, and pressed into a pellet that was put into a tube furnace and heat-treated at different temperatures for 24 h under flowing O_2. For the multiple sintering sample, the product powders were ground, pressed and heat-treated again under the same process. The Er211 powders were synthesized by a similar process with the sintering parameters of 1000 °C for 24 h under the flow Ar. The oxygen annealing process used the Er123 sample with the highest purity. The as-synthesized powders were ground and pressed to pellet. The pellet was put into a tube furnace and heat-treated at 500 °C for 24 h or 48 h under flowing O_2 with a large flow rate. The commercial Gd123 powder is coming from Shanghai Superconductor Technology Co., Ltd.

Sample Characterization. The phase analysis of the samples was characterized by Cu Kα x-ray diffraction (XRD, Burker D8), and the grain sizes were calculated by using the Debye–Scherrer formula. Microstructures were observed by scanning electron microscopy (SEM, FEI Quanta 450) with Energy Dispersive X-Ray Spectroscopy (EDX). The melt temperature was measured by differential thermal analysis (DTA-TGA, TA Instruments 2960) at a heating rate of 10 °C/min in the Air. Magnetic measurements were performed at 5K and 77 K in a vibrating sample magnetometer (VSM) using a Physical Properties Measurement System (PPMS) from Quantum Design Ltd. under an applied field up to 7 T. Field-cooled (FC) and the zero-field-cooled (ZFC) curve was measured in an applied field of 10 mT. Magnetization values were determined from the measured magnetic moment using the sample mass and nominal density ρ = (Gd123: 6.384 g/cm^3, Er123: 7.152 g/cm^3) to calculate the actual volume of material present. The critical current densities, J_c (in Am^{-2}), of the samples were calculated by applying a standard Bean model expression for spherical grains to magnetic hysteresis loops through the formula:

$$\Delta M = \frac{1}{3}\frac{w}{2} J_c\left(1 - \frac{w}{3l}\right)$$

where ΔM (in Am^{-1}) is the vertical width of the magnetization loop and $l \geq w \gg t$ (in m) are the dimensions of individual plate-like crystallites in the samples. This formula was derived in Reference [27] for a collection of randomly-oriented thin platelets of an anisotropic superconductor such as REBCO, and yields an estimate of the ab-plane J_c for fields applied parallel to c typically accessed in transport measurements while incorporating considerations relating to the anisotropy of the superconductor, the geometry of the crystallites, their respective orientations to the applied field and demagnetization effects. In the case of the commercial Gd123 powder, the standard Bean model expression for spherical grains $\Delta M = d/3 J_c$ has been used, where d is the grain diameter (taken as 5 μm).

3. Results

Figure 1a shows the XRD results of the sample sintered at different temperatures for 24 h. All the samples consist of Er123 as a domain phase with impurities of Er_2BaCuO_5 (Er211) and $BaCuO_2$. It shows that the sintering temperature has an apparent effect on the phase formation. As shown in Figure 1b, along with the temperature raising to 920 °C, the peak intensity of Er123 kept increasing, at the same time the Er211 peak became weaker. When the temperature reaches 930 °C, such a trend reversed. The grain size calculated from XRD results shows the same evolution trend with the Er123 peak intensity.

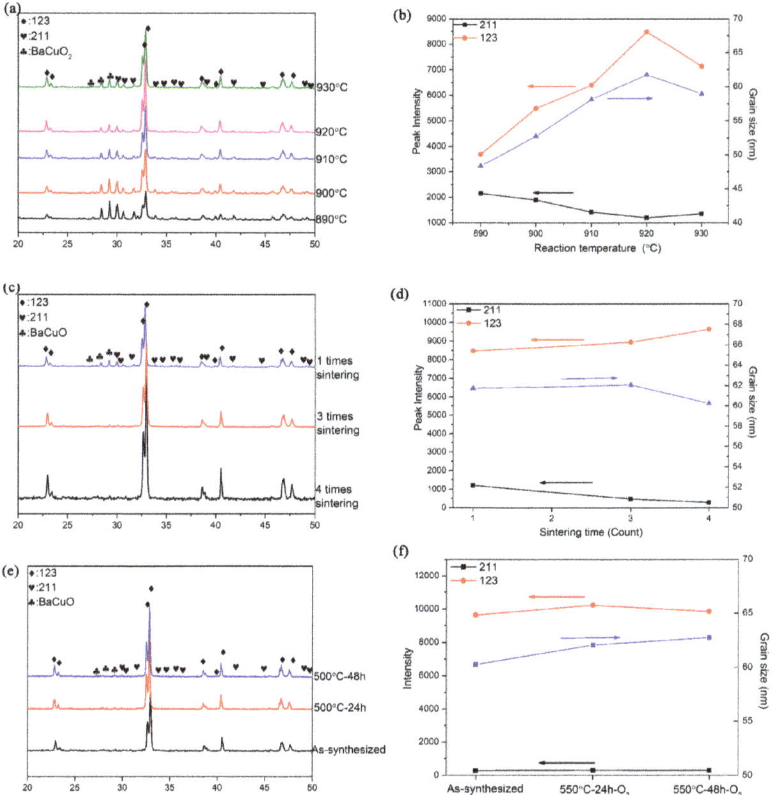

Figure 1. (**a**) phase formation of Er123 sintered by different temperature for 24 h; (**b**) Peak intensity and grain size from the XRD results of Er123 sintered by different temperature for 24 h; (**c**) phase formation of Er123 sintered at 920 °C for 24 h by multiple times; (**d**) Peak intensity and grain size from the XRD results of Er123 sintered at 920 °C for 24 by multiple times; (**e**) phase formation of Er123 with oxygen annealing at 500 °C for 24~48 h; (**f**) Peak intensity and grain size from the XRD results of Er123 with oxygen annealing at 500 °C for 24~48 h.

Multiple heat treatment is the conventional method to improve the reaction in the solid reaction method. The multiple heat treatment under 920 °C has been attempted, as shown in Figure 1c,d. The impurity phase such as Er211 and $BaCuO_2$ nearly disappeared after 4 times sintering. Additionally, the grain size does not have significant changing along with the multiple heat treatment.

The purpose of oxygen annealing on the REBCO powder is to compensate for the oxygen deficiency brought by the high temperature heat treatment, the annealing at 500 °C for 24 and 48 h was also tried. As shown in Figure 1e,f, the change of phase formation and grain size before and after oxygen annealing is negligible, which both have high purity Er123 as the dominant phase.

Figure 2 shows the SEM images of the sample sintered at different temperatures for 24 h. When the temperature was 890 °C, there are plenty of small round particles embedded inside big Er123 particles, as pointed by the red arrow. The EDX results showed that these small particles are the Er211 phase. As the temperature rose, these small round particles gradually disappeared. When the temperature reaches the range of 910~920 °C, only the morphology of big plate-like particles can be found. However, the big plate-like particles started to disassemble at a temperature of 930 °C. This is due to the reaction between the Er123 and CuO to form Er211 and liquid phase [28]. Combined with the XRD results, we think it is due to the decomposition of Er123 to Er211 at the excessive temperature. However, we did not find the small round Er211 particles as in the sample of 890 °C. We randomly chose 50 particles and measured the particle size of the plate-like Er123. The average particles size of the Er123 in the sample of 900~920 °C are 5 μm × 5 μm × 1 μm. The particle size has a small change in the sample of 890 °C and 930 °C, due to the existence of Er211 impurities. However, the same value will be used in the calculation of the Bean model.

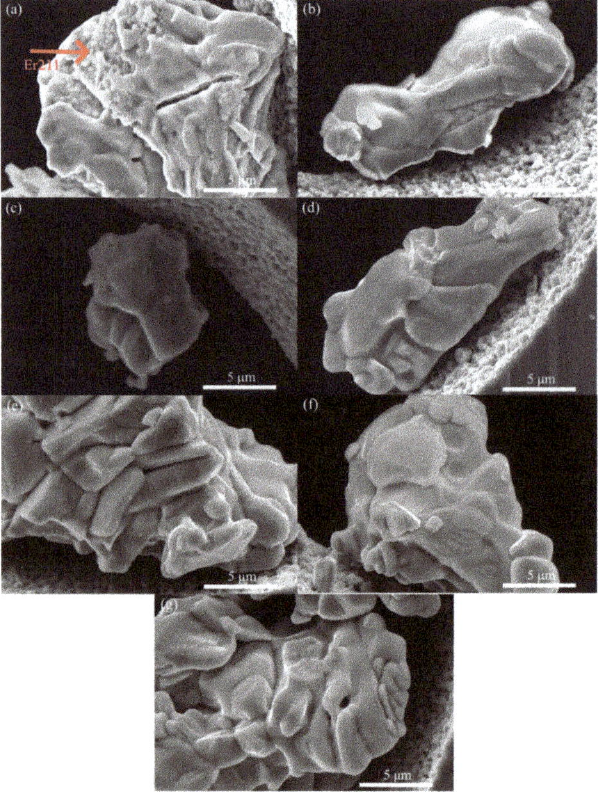

Figure 2. The SEM images of Er123 powder after different sintering. (**a**): 890 °C for 24 h; (**b**): 900 °C for 24 h; (**c**): 910 °C for 24 h; (**d**): 920 °C for 24 h; (**e**): 920 °C for 24 h (three times); (**f**): 920 °C for 24 h (four times); (**g**): 930 °C for 24 h.

We did multiple heat treatment on the sample of 920 °C. Although the phase purity was further improved, there was no apparent change in the microstructure, as shown in Figure 2e,f. The grain size also did not change after multiple heat treatment. The microstructure and particle size of the sample with four times sintering at 920 °C for 24 h and oxygen annealing at 500 °C for 48 h are nearly the same as the sample with only one-time sintering at 920 °C for 24 h. It indicates that the Er123 particle with a size around 5μm is very stable, which is hard to connect together under more heat treatment.

Figure 3 shows the superconducting properties of the sample sintered with different parameters. As shown in Figure 3a, under different sintering temperatures, the $T_{c\,onset}$ value is nearly the same—around 91 K—but there is an apparent difference in the magnetization value. In both the ZFC and FC curves, the magnetization value kept increasing and reached the maximum at 920 °C. The ZFC value shows how many superconductor phases in the sample. That is the reason the magnetization value in the ZFC curve has exactly the same trend as the peak intensity of Er123. The FC magnetization value is associated with the Meissner fraction, and is much lower in magnitude for both samples, indicating flux trapping within the grains. The increasing FC magnetization value indicates the poor pinning in all the samples. All the J_c curves of the sample with different sintering temperatures are parallel to each other at 5 K up to 7 T, as shown in Figure 3b. The sample of 920 °C has the highest J_c value, which is due to the highest Er123 phase amount. The 211 phase is the most conventional pinning center in the REBCO bulks. However, although there is a visible Er211 phase in both the sample of 890 °C and 930 °C, we did not find any better field performance in these two samples.

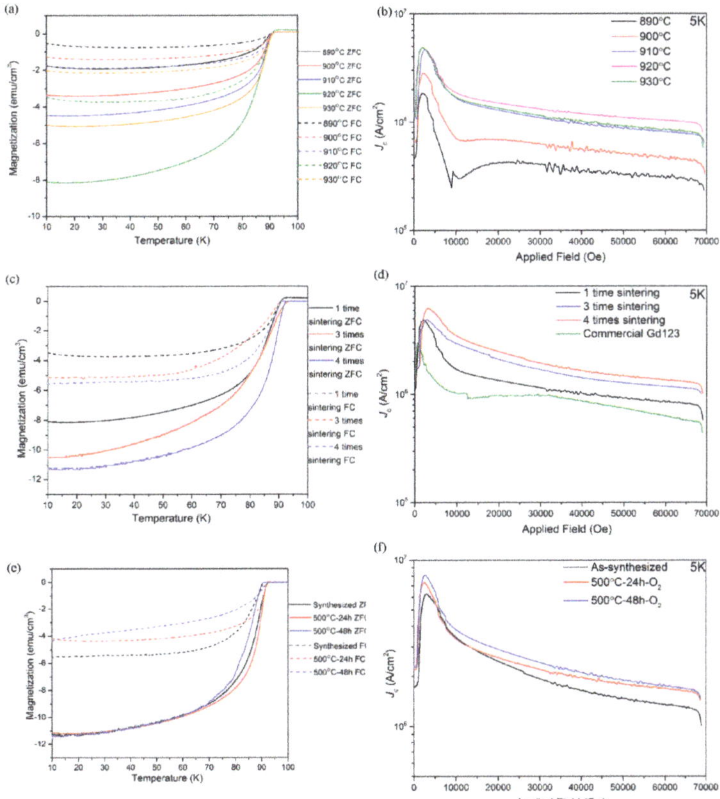

Figure 3. (**a**) ZFC and FC curves of the Er123 sintered by different temperature for 24 h; (**b**) J_c vs B curve at 5 K of the Er123 sintered by different temperature for 24 h; (**c**) ZFC and FC curves of the Er123 sintered at 920 °C for 24 h by multiple times; (**d**) J_c vs B curve at 5 K of the Er123 sintered at 920 °C for 24 h by multiple times; (**e**) ZFC and FC curves of the Er123 with oxygen annealing at 500 °C for 24~48 h; (**f**) J_c vs B curve at 5 K of the Er123 with oxygen annealing at 500 °C for 24~48 h.

As shown above, the multiple heat treatment at 920 °C can further improve the phase purity. The enhancement of ZFC magnetization value and J_c value also proved such results, as shown in Figure 3c,d. Moreover, the J_c value at 5K is higher than the commercial Gd123 powder at the entire 0~7 T. Since the Gd123 powder is the exact one that Shanghai Superconductor Technology Co., Ltd. used to fabricate Gd123 coated conductor, this result indicates that the Er123 has similar intra-grain supercurrent with Gd123 which is used to fabricate commercial coated conductor. If the excellent texture of Er123 can be obtained between Gd123 commercial coated conductor by exquisitely designing the heat treatment process, we also expect the same level inter grain supercurrent of Er123 to Gd123, which finally provides a possibility to fabricate superconducting joint with high current capability by using Er123 as superconducting solder.

Although the phase formation and microstructure did not change after oxygen annealing, the superconducting properties showed some difference, as shown in Figure 3e,f. After annealing at 500 °C under O_2, the ZFC magnetization values did not have further improvement, indicated that although sintering at 920 °C could cause oxygen deficiency, the cooling underflow oxygen can basically compensate the deficiency. However, the FC magnetization values slightly decreased with the prolonging of the annealing time at 500 °C. The J_c curves also tell a new story: the high filed performance was mildly improved in the sample with oxygen annealing. Both results show that oxygen annealing improves the pinning in the Er123 powder.

4. Discussion

From the results above, it can be found that there is a reversible reaction between Er123 and Er211 phases. At lower temperatures, the Er211 can react with $BaCuO_2$ form Er123, moreover, the Er123 can also decompose to Er211. This is the reason the Er123 intensity has a convex shape with the increasing temperature, and the optimized temperature windows on the Er123 formation is pretty narrow. These Er211 seems to agglomerate to particles of few micrometers and embedded into the large Er123 particles. According to the superconducting properties, especially the high field performance of J_c value, such embedded Er211 particles could not act as the pinning center but only degrade the T_c value. After multiple heat treatments, the impurity phase such as Er211 can be eliminated; however, the poor pinning problem emerges. Therefore, the controllable Er211 particles which can act as pinning centering are essential to the Er123 as the superconducting solder. We successfully synthesized the high purity Er211 phase with a particle size of around 10 micrometers, as shown in Supplementary Materials Figure S1. However, simply mixing such Er211 will not bring the pinning center, and further refining work is necessary.

During recent decades, research on the melt texture synthesis of REBCO single domain bulk, multiple additives has proved that it is a benefit to the superconducting properties of the final products. Typically, the Ag or Ag_2O is added to improve the connectivity and decrease the melt temperature of REBCO [29–32], which brings lower heat treatment temperature. Pt is also a common additive, which can inhibit the growth of the 211 phase [33,34]. The large 211 particles can lead to an inhomogeneous reaction. All these beneficial dopants will be introduced in the fabrication of a superconducting joint by using Er123 as a superconducting solder. Therefore, the effect of such additives on the Er123 melt temperature is critical, which directly determines the heat treatment process of fabrication of the superconducting joint.

After obtaining Er123 powder with high purity and high supercurrent capability, the feasibility of Er123 as a superconducting solder is discussed here. Since the REBCO need the biaxial texture to carry high supercurrent, the only method of fabricating a superconducting joint between REBCO coated conductor is using REBCO materials to form a texture joint. In the most conventional method, the REBCO powder of low melt temperature is added as a superconducting solder between the REBCO coated conductor. During the heat treatment, the solder REBCO melts but the REBCO coated conductor keeps inert. After cooling at a slow rate, the solder REBCO grows to texture form based on the REBCO coated conductor. Therefore, the low melting temperature of superconducting solder is critical.

Figure 4 shows the DTA results of the Er123 powder and Gd123 powder with a different dopant in Air. It is found that the melt temperature of Er 123 decreased about 30 °C by adding Ag and Ag_2O, but no such effect is found in the curve of Pt adding. The melt temperature decreasing by adding Ag and Ag_2O is very useful, which may bring more safety margins when designing the heat treatment process of the superconducting joint.

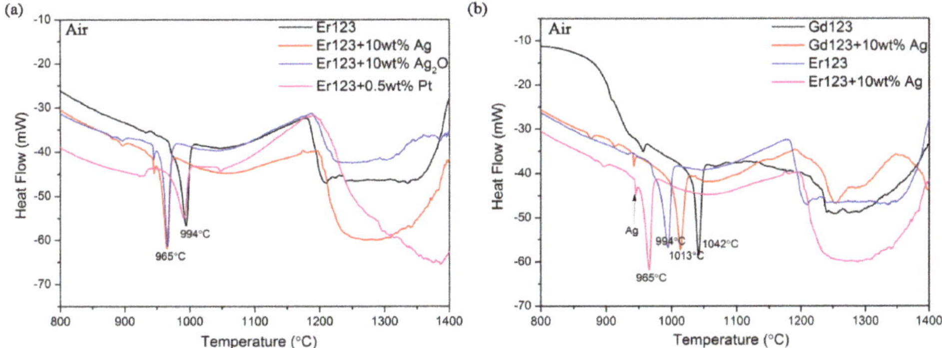

Figure 4. (a) The DTA results of Er123 powder with different dopants; (b) The DTA results of Er123 and Gd123 powder with and without Ag doping.

However, according to our previous study, the Ag can decrease the melt temperature of REBCO by just contacting instead of homogenous mixing [35]. In the report, we found that the Ag can decrease the melt temperature of Yb123 to the same level as the Yb123 powder mixed with Ag. So if the Ag is added to Er123 during the fabrication of the superconducting joint between the G123 coated conductor, the effect of Ag on the melt temperature of Gd123 should be considered. Figure 4b shows the DTA results of the Er123 and Gd123 powder with and without Ag adding. It is found that the melt temperatures of both theGd123 and Er123 were decreased by the same level (about 29 °C) with Ag adding. So the heat treatment temperature range moved from 994~1042 °C to 965~1013 °C without any expansion. For safety reasons, the maximum heat treatment temperature should set below 1013 °C instead of 1042 °C, in case of the melting of Gd123 under the effect of Ag.

5. Conclusions

The synthesis parameters of Er123 powder were systematically optimized. The optimization temperature was 920 °C; both higher or lower temperatures brought small Er211 particles embedded in Er123 plate-like particles. The Er211 phase is a kind of barrier of the flow superconducting current but cannot act as a pinning center. After multiple sintering processes, high purity Er123 powder with as good superconducting properties as Gd123 coated conductor was acquired. The extra oxygen annealing at 500 °C is not necessary for this synthesis method. The melt temperature of Er123 and Gd123 with different dopants were also investigated. After adding Ag or Ag_2O, a feasible operating temperature range (965~1013 °C) is uncovered, at which the superconducting solder Er123 can melt but the Gd123 in coated conductor remains inert. This shows the important first step to obtain a superconducting joint with high current capability.

Supplementary Materials: The following are available online at http://www.mdpi.com/2073-4352/9/10/492/s1, Figure S1: The XRD results of Er211 powders.

Author Contributions: Z.Z. contributed to the main part of this article including investigation, analysis and original writing. The L.W., J.L. and Q.W. contributed to the supervision of the investigation and writing—review and editing.

Funding: Z. Zhang was supported by the National Natural Science Foundation of China (51702316). L. Wang was supported by the Beijing Natural Science Foundation (3184061). Liu and Q. Wang was supported by the National

Natural Science Foundation of China (51777205 and 11745005) and International partnership program of Chinese Academy of Sciences (182111KYSB20170039).

Acknowledgments: We acknowledge Shanghai Superconductor Technology Co., Ltd. to provide Gd123 powder.

Conflicts of Interest: The authors declare no conflicts of interest.

References

1. Iwasa, Y.; Bascnan, J.; Hahn, S.; Yao, W. High-temperature superconductors for NMR/MRI magnets: Opportunities and challenges. *Prog. Supercond. Cryog.* **2009**, *11*, 1–7.

2. Yanagisawa, Y.; Piao, R.; Iguchi, S.; Nakagome, H.; Takao, T.; Kominato, K.; Hamada, M.; Matsumoto, S.; Suematsu, H.; Jin, X. Operation of a 400MHz NMR magnet using a (RE:Rare Earth)$Ba_2Cu_3O_{7-x}$ high-temperature superconducting coil: Towards an ultra-compact super-high field NMR spectrometer operated beyond 1GHz. *J. Magn. Reson.* **2014**, *249*, 38–48. [CrossRef] [PubMed]

3. Iwasa, Y.; Bascunan, J.; Hahn, S.; Voccio, J.; Kim, Y.; Lecrevisse, T.; Song, J.; Kajikawa, K. A High-Resolution 1.3-GHz/54-mm LTS/HTS NMR Magnet. *IEEE Trans. Appl. Supercond.* **2015**, *25*, 4301205. [CrossRef]

4. Tosaka, T.; Miyazaki, H.; Iwai, S.; Otani, Y.; Takahashi, M.; Tasaki, K.; Nomura, S.; Kurusu, T.; Ueda, H.; Noguchi, S.; et al. R&D Project on HTS Magnets for Ultra high-Field MRI Systems. *IEEE Trans. Appl. Supercond.* **2016**, *26*, 4402505.

5. Yokoyama, S.; Lee, J.; Imura, T.; Matsuda, T.; Eguchi, R.; Inoue, T.; Nagahiro, T.; Tanabe, H.; Sato, S.; Daikoku, A.; et al. Research and Development of the High Stable Magnetic Field ReBCO Coil System Fundamental Technology for MRI. *IEEE Trans. Appl. Supercond.* **2017**, *27*, 4400604. [CrossRef]

6. Piao, R.; Iguchi, S.; Hamada, M.; Matsumoto, S.; Suematsu, H.; Saito, A.T.; Li, J.; Nakagome, H.; Takao, T.; Takahashi, M.; et al. High resolution NMR measurements using a 400 MHz NMR with an (RE)$Ba_2Cu_3O_{7-x}$ high-temperature superconducting inner coil: Towards a compact super-high-field NMR. *J. Magn. Reson.* **2016**, *263*, 164–171. [CrossRef] [PubMed]

7. Swenson, C.A.; Markiewicz, W.D. Persistent joint development for high field NMR. *IEEE Trans. Appl. Supercond.* **1999**, *9*, 185–188. [CrossRef]

8. Kim, Y.; Bascunan, J.; Lecrevisse, T.; Hahn, S.; Voccio, J.; Park, D.K.; Iwasa, Y. YBCO and Bi2223 coils for high field LTS/HTS NMR magnets: HTS-HTS joint resistivity. *IEEE Trans. Appl. Supercond.* **2013**, *23*, 6800704.

9. Baldan, C.A.; Oliveira, U.R.; Bernardes, A.A.; Oliveira, V.P.; Shigue, C.Y.; Ruppert, E. Electrical and superconducting properties in lap joints for YBCO tapes. *J. Supercond. Nov. Magn.* **2013**, *26*, 2089–2092. [CrossRef]

10. Walsh, R.P.; McRae, D.; Markiewicz, W.D.; Lu, J.; Toplosky, V.J. The 77-K stress and strain dependence of the critical current of YBCO coated conductors and lap joints. *IEEE Trans. Appl. Supercond.* **2012**, *22*, 8400406. [CrossRef]

11. Park, Y.; Lee, M.; Ann, H.; Choi, Y.K.; Lee, H. A superconducting joint for $GdBa_2Cu_3O_{7-\delta}$ coated conductors. *NPG Asia Mater.* **2014**, *6*, e98. [CrossRef]

12. Jin, X.Z.; Yanagisawa, Y.; Maeda, H.; Takano, Y. Development of a superconducting joint between a GdBa2Cu3O7-δ-coated conductor and $YBa_2Cu_3O_{7-\delta}$ bulk: Towards a superconducting joint between RE (Rare Earth) $Ba_2Cu_3O_{7-\delta}$-coated conductors. *Supercond. Sci. Technol.* **2015**, *28*, 075010. [CrossRef]

13. Jin, X.Z.; Yanagisawa, Y.; Maeda, H. Measurement of Persistent Current in a Gd123 Coil with a Superconducting Joint Fabricated by the CJMB Method. *IEEE Trans. Appl. Supercond.* **2018**, *28*, 4602604. [CrossRef]

14. Noudem, J.G.; Reddy, E.S.; Tarka, M.; Noe, M.; Schmitz, G.J. Melt-texture joining of $YBa_2Cu_3O_y$ bulks. *Supercond. Sci. Technol.* **2001**, *14*, 363–370. [CrossRef]

15. Mukhopadhyay, S.M.; Mahadev, N.; Sengupta, S. Microstructural and spectroscopic analyses of a strongly-linked joint formed in a superconductor. *Physica C* **2000**, *329*, 95–101. [CrossRef]

16. Zheng, H.; Jiang, M.; Nikolova, R.; Welp, U.; Paulikas, A.P.; Huang, Y.; Crabtree, G.W.; Veal, B.W.; Claus, H. High critical current "weld" joints in textured $YBa_2Cu_3O_x$. *Physica C* **1999**, *322*, 1–8. [CrossRef]

17. Schmitz, G.J.; Tigges, A.; Schmidt, J.C. Microstructural aspects of joining superconductive components using (RE)$Ba_2Cu_3O_{7-x}$ solder. *Supercond. Sci. Technol.* **1998**, *11*, 73–75. [CrossRef]

18. Azoulay, J. YBCO thin film evaporation on as-deposited silver film on MgO. *Physica C* **1999**, *324*, 187–192. [CrossRef]

19. Shimoyama, J.I.; Horii, S.; Otzschi, K.; Kishio, K. How to optimize critical current performance of RE123 materials by controlling oxygen content. *Mater. Res. Soc. Symp. Proc.* **2002**, *689*, 265–269. [CrossRef]

20. Horii, S.; Ichinose, A.; Mukaida, M.; Matsumoto, K.; Ohazama, T.; Yoshida, Y.; Shimoyama, J.I.; Kishio, K. Enhancement of Critical Current Density in $ErBa_2Cu_3O_y$ Thin Films by Post-Annealing. *Jpn. J. Appl. Phys.* **2004**, *43*, L1223–L1225. [CrossRef]

21. Iida, K.; Yoshioka, J.; Sakai, N.; Murakami, M. Superconducting joint of Y–Ba–Cu–O superconductors using Er–Ba–Cu–O solder. *Physica C* **2002**, *370*, 53–58. [CrossRef]

22. Nakashima, T.; Maruyama, T.; Honzumi, M.; Horii, S.; Shimoyama, J.; Kishio, K. Relationship between Critical Current Properties and Microstructure of Er123 Melt-Solidified Bulks. *IEEE Trans. Appl. Supercond.* **2005**, *15*, 3176–3179. [CrossRef]

23. Kita, R.; Hosoya, N.; Otawa, N.; Kawabata, S.; Nakamura, T.; Miura, O.; Mukaida, M.; Yamada, K.; Ichinose, A.; Matsumoto, K.; et al. Effects of RE_2O_3 (RE = Tm, Sc, Yb) addition on the superconducting properties of $ErBa_2Cu_3O_y$. *Physica C* **2009**, *469*, 1157–1160. [CrossRef]

24. Yoshioka, J.; Iida, K.; Negichi, T.; Sakai, N.; Noto, K.; Murakami, M. Joining Y123 bulk superconductors using Yb–Ba–Cu–O and Er–Ba–Cu–O solders. *Supercond. Sci. Technol.* **2002**, *15*, 712–716. [CrossRef]

25. Iida, K.; Kono, T.; Kaneko, T.; Katagiri, K.; Sakai, N.; Murakami, M.; Koshizuka, N. Joining of Y–Ba–Cu–O/Ag bulk superconductors using Er–Ba–Cu–O/Ag solder. *Supercond. Sci. Technol.* **2004**, *17*, S46–S50. [CrossRef]

26. Iida, K.; Yoshioka, J.; Negichi, T.; Noto, K.; Sakai, N.; Murakami, M. Strong coupled joint for Y–Ba–Cu–O superconductors using a sintered Er–Ba–Cu–O solder. *Physica C* **2002**, *378–381*, 622–626. [CrossRef]

27. Peterson, R.L. Magnetization of Anisotropic Superconducting Grains. *J. Appl. Phys.* **1990**, *67*, 6930–6933. [CrossRef]

28. MacManus-Driscol, J.L. Materials Chemistry and Thermodynamics of $REBa_2Cu_3O_{7-x}$. *Adv. Mater.* **1997**, *9*, 457–473. [CrossRef]

29. Diko, P.; Fuchs, G.; Krabbes, G. Influence of silver addition on craking in melt-grown YBCO. *Physica C* **2001**, *363*, 60–66. [CrossRef]

30. Ates, A.; Yanmaz, E. The effects of Ag addition and magnetic field on melt-processed $YBa_2Cu_3O_x$ superconductors. *J. Alloys Comp.* **1998**, *279*, 220–228. [CrossRef]

31. Cakir, B.; Aydiner, A. Structural and magnetic properties of the ring shaped 40 wt% Y211 added TSMG Y123 bulk superconductors welded by Ag2O added MPMG YBCO solder material. *J. Mater. Sci. Mater. Electron.* **2017**, *28*, 17098–17106. [CrossRef]

32. Diko, P.; Krabbes, G.; Wende, C. Influence of Ag addition on crystallization and microstructure of melt-grown single-grain $YBa_2Cu_3O_7$ bulk superconductors. *Supercond. Sci. Technol.* **2001**, *14*, 486–495. [CrossRef]

33. Manton, S.J.; Beduz, C.; Yang, Y.F. Rejoining of Single Grain Melt Textured Bulk YBazCu30,-x I. *IEEE Trans. Appl. Supercond.* **1999**, *9*, 2089–2092. [CrossRef]

34. Zhou, L.; Chen, S.K.; Wang, K.G.; Wu, X.Z.; Zhang, P.X.; Feng, Y.; Wen, H.H.; Li, S.L. Preparation of enhanced Jc YBCO bulks by powder melting process with a combination of submicron 211 precursor and Pt addition. *Physica C* **2002**, *371*, 62–68. [CrossRef]

35. Zhang, Z.; Jiang, J.; Tian, H.; Wang, Q.; Larbalestier, D.; Hellstrom, E. Investigation of the melt-growth process of $YbBa_2Cu_3O_{7-\delta}$ powder in Ag-sheathed tapes. *CrystEngComm* **2019**, *21*, 1369–1377. [CrossRef]

© 2019 by the authors. Licensee MDPI, Basel, Switzerland. This article is an open access article distributed under the terms and conditions of the Creative Commons Attribution (CC BY) license (http://creativecommons.org/licenses/by/4.0/).

Review

Highly-Tunable Crystal Structure and Physical Properties in FeSe-Based Superconductors

Kaiyao Zhou [1,2]**, Junjie Wang** [1,2]**, Yanpeng Song** [1,2]**, Liwei Guo** [1] **and Jian-gang Guo** [1,3,]*

[1] Beijing National Laboratory for Condensed Matter Physics, Institute of Physics, Chinese Academy of Sciences, Beijing 100190, China; kaiyaozhou@hotmail.com (K.Z.); wangjunjie1219@hotmail.com (J.W.); syp2226340310@hotmail.com (Y.S.); lwguo@aphy.iphy.ac.cn (L.G.)

[2] University of Chinese Academy of Sciences, Beijing 100049, China

[3] Songshan Lake Materials Laboratory, Dongguan, Guangdong 523808, China

* Correspondence: jgguo@iphy.ac.cn

Received: 12 September 2019; Accepted: 22 October 2019; Published: 25 October 2019

Abstract: Here, crystal structure, electronic structure, chemical substitution, pressure-dependent superconductivity, and thickness-dependent properties in FeSe-based superconductors are systemically reviewed. First, the superconductivity versus chemical substitution is reviewed, where the doping at Fe or Se sites induces different effects on the superconducting critical temperature (T_c). Meanwhile, the application of high pressure is extremely effective in enhancing T_c and simultaneously induces magnetism. Second, the intercalated-FeSe superconductors exhibit higher T_c from 30 to 46 K. Such an enhancement is mainly caused by the charge transfer from the intercalated organic and inorganic layer. Finally, the highest T_c emerging in single-unit-cell FeSe on the $SrTiO_3$ substrate is discussed, where electron-phonon coupling between FeSe and the substrate could enhance T_c to as high as 65 K or 100 K. The step-wise increment of T_c indicates that the synergic effect of carrier doping and electron-phonon coupling plays a critical role in tuning the electronic structure and superconductivity in FeSe-based superconductors.

Keywords: FeSe; superconductivity; high pressure; chemical intercalation; interfacial coupling

---

1. Introduction

After discovering FeAs-based superconductors in 2008 [1–4], a structurally-simple binary FeSe superconductor with a superconducting critical temperature (T_c) of 8 K was quickly established by Hsu et al. [5] This brought researchers into a fresh new field of iron-based superconductors. The binary compound is composed of a neutral FeSe layer that is composed of a $FeSe_4$ tetrahedra along the c-axis, and thus the interlayer coupling is weak van der Waals forces. It was found, initially, that the chemical substitution of Se by S and Te mildly enhanced the T_c to 10−15 K. However, tiny amounts of doping (3%) in the Fe site by Cu and Co strikingly suppress the T_c. Such features are totally different from those of FeAs-based superconductors, where the Co/Ni doping effectively induces superconductivity (SC) accompanying many properties like linear-T resistivity, a quantum critical point, and magnetic fluctuations [6–8]. Furthermore, there is no long-range magnetic order in FeSe, and the interplay between magnetism and SC is lacking, thus its superconducting mechanism seems to be simpler than that of FeAs-based superconductors. On the other hand, it is hard to grow large-sized single crystals of FeSe by the conventional flux-method because superconducting FeSe is a metastable phase that only exists in the temperature range of 673 K−973 K. The intrinsic properties like gap symmetry and the Fermi surface are hence inaccessible. These disadvantages, to some extent, hinder the investigations of FeSe-based superconductors.

Later, a technique of growing high-quality FeSe single crystals was found, and multiple characterizations like magneto-transport, in situ pressure, and inelastic neutron scattering revealed inherent properties for SC. More importantly, in 2010, the intercalation between the FeSe layer by large-sized alkali metals was successfully realized, forming a series of superconductors $A_xFe_{2-y}Se_2$ (A = K, Cs, Rb, Tl) with a T_c of 30 K. This discovery opened a broader field in the physics and chemistry community. Subsequent reports of alkali metal, and ammonia co-intercalated FeSe, inorganic molecule intercalation, and monolayer FeSe on a $SrTiO_3$ substrate continuously enhanced the T_c up to 65 K or even 100 K, making the FeSe-based compound one of the most fascinating fields in the search for a high T_c superconductor. Apart from the enhancement of T_c, the angle resolve photoemission spectrum (ARPES) revealed that the Fermi surface in such high T_c superconductors share a common feature (i.e., only two electron pockets surviving at the M point). A unique Fermi surface possibly implies a different pairing mechanism for Cooper pairs rather than the Fermi nesting scenario between electron-hole pockets in the well-documented FeAs-based superconductor [9,10].

In the last 11 years, FeSe-based superconductors have exhibited huge advancements and prosperous developments, and we feel that it is necessary to summarize the significant progress made at each important step. In this paper, we will review the crystal structure, the electronic structure, and the superconducting phase diagram in diverse FeSe-based superconductors. The evolutions of SC in binary FeSe under chemical substitution and physical pressure are discussed in first part. Second, the FeSe-based intercalates through inorganic and organic molecules are summarized, where both the interlayer spacing between FeSe layers and the charge transfer should determine the T_c. Finally, the substrate effect, the doping carrier, and the interfacial coupling of high-T_c SC in thin FeSe film are discussed. We hope that this review can deepen the understanding of structural evolution and superconductivity in FeSe-based superconductors to motivate the exploration of new superconducting materials with higher T_c.

2. Superconductivity in FeSe

2.1. Structure and Stoichiometric of $Fe_{1+x}Se$

Binary FeSe with T_c = 8 K was discovered in 2008. It is regarded as a less-toxic superconductor when compared with the FeAs-based superconductors. The structure of FeSe is the PbO-type, which is composed of FeSe layers that are formed by edge-shared $FeSe_4$ tetrahedra. The onset of T_c is 8.5 K. Rietveld refinements of powder synchrotron diffraction (BL12B2 beamline in Spring8) yield a = 0.37693(1) nm and c = 0.54861(2) nm for $FeSe_{0.82}$, and a = 0.37676(2) nm and c = 0. 54847(1) nm for $FeSe_{0.88}$. By cooling below the phase transition temperature (T_s) of 90 K, the (220) peak of PXRD splits into (200) and (020), indicating a structural transition from tetragonal $P4/nmm$ to triclinic P-1, as shown in Figure 1. After that, Margadonna et al. carefully resolved the low-temperature crystal structures for $FeSe_{0.92}$ [11]. Both high-resolution synchrotron x-ray diffraction and powder neutron diffraction (NPD) were carried out at different temperatures. They found that it indeed experienced a structural transition below the T_s, but the low-temperature phase showed an orthorhombic phase (space group Cmma), rather than triclinic symmetry, as seen in Figure 2. This transition was later called a nematic transition, which possibly correlates with the SC evidenced by multiple characterizations [12,13].

Pomjakushina et al. used two different methods to synthesize the $FeSe_{1-x}$ powder [14], and found that the superconducting phase existed in a very narrow range of Se content ($FeSe_{0.974\pm0.005}$). Williams et al. carefully identified the role of Se-deficiency on tuning the structural transition and SC [15]. They proved experimentally that the correct formula of superconducting samples was $FeSe_{0.99}$. The Se vacancies may not be necessary for inducing SC. They also determined that the SC was extremely sensitive to the content and disorder of Fe. After introducing 3% excess Fe, the SC was totally suppressed. Additionally, the magnetic order was not observed down to 5 K, in contrast to the antiferromagnetic (AFM) ground state in the FeAs-based superconductors. The $Fe_{1.01}Se$ compound exhibits the superstructure below the T_s in the form of a $\sqrt{2} \times \sqrt{2} \times 1$ supercell, consistent with the previous report [16]. Moreover, in the compound $Fe_{1.03}Se$, the structural transition disappears, which means that excess Fe can break the long-range coherence of the structural unit. Recently, more details on the superstructure and SC were revealed in vacancy-ordered $Fe_{1-x}Se$. Three kinds of superstructures, $\sqrt{2} \times \sqrt{2} \times 1$, $\sqrt{5} \times \sqrt{5} \times 1$, and $\sqrt{10} \times \sqrt{10} \times 1$, were identified in Fe_3Se_4, Fe_4Se_5, and superconducting Fe_9Se_{10}, respectively [17]. Among them, the Fe_4Se_5 compound is an AFM insulator with a small effective magnetic moment (0.003 μ_B). These authors claimed that the Fe vacancies could be effectively tuned, and increasing the content of Fe gradually led to the so-called parent compound Fe_4Se_5 into a superconducting state. Such behavior is analogous to the superconducting phase diagram in cuprates, where the parent compounds are AFM insulators and excess hole or electron doping can induce SC [18].

Regarding the single crystal of FeSe, Mok et al. could grow it through the high-temperature method using flux KCl [19]. As shown in Figure 3a, single crystals 2–3 mm in width and 0.1–0.3 mm in thickness were obtained. Only small amounts of the hexagonal FeSe impurity phase was detected according to the x-ray diffraction analysis. This work further proved that post-annealing at 673 K was useful to improve the quality of single crystals. Chareev et al. successfully grew high-quality FeSe single crystals by the chemical vapor deposition method. The atomic ratio of Fe:Se = 1:0.96 ± 0.02 and the maximal size was $4 \times 4 \times 0.1$ mm^3. The layered feature can clearly be observed in the Figure 3b. The $T_c{}^{onset}$ of this crystal was 9.4 K [20]. Following the improvement of single crystals, several groups quickly determined the intrinsic SC of FeSe. Lei et al. presented the nearly isotropic upper critical field $[H_{c2}(0)]$ with magnetic field parallel and perpendicular to (101) face (see Figure 4a,b), and deduced a large Ginzburg-Landau parameter of $\kappa \sim 72.3(2)$ [21]. The critical current density (J_c) was determined to be 2.2×10^4 A/cm^2 by the Bean model. Furthermore, these authors analyzed the Hall effect of FeSe and confirmed that the dominant carriers were hole-type at high magnetic fields. The electron-type carriers made a larger contribution at low magnetic field and low temperature, as shown in Figure 4c [22]. The symmetry of the superconducting energy gap was investigated by Lin et al. who demonstrated the coexistence of the isotropic *s*-wave and extended anisotropic *s*-wave gap with the magnitude of 1.33 meV and 1.13 meV in FeSe [23]. A high electron-boson coupling constant, 1.55, implies that the electron is not only coupled with phonons, but also with other "glues" associated with spin fluctuations. Subsequently, Abdel-Hafiez et al. measured the temperature-dependent London penetration depth, which was fitted by either a two s-wave-like model or a single anisotropic gap [24]. They claimed that the superconducting energy gap is nodeless. The fitting curves of a two-gap model with anisotropic *s*-wave and *d*-wave models are shown in Figure 4d.

Density functional calculations demonstrated that the electronic structure of FeSe was very similar to those of the FeAs-based superconductors [25]. In the first Brillouin zone, there are heavy hole-cylinders and lighter electron-pockets near the zone center (Γ-point) and corner (M-point), respectively, implying typical spin-density wave instability due to Fermi surface nesting. In 2014, Shimojima et al. investigated the temperature-dependent band dispersion of Fe $3d_{xz}$ and $3d_{yz}$ orbitals at the M point from 50 K to 110 K. The results showed that the removal of degeneracy in the $3d_{xz}/3d_{yz}$ bands occurred at temperatures close to the T_s, which may be the orbital origin of structural transition [26]. Later, Zhang et al. traced the temperature dependence of band dispersion at the M point and the Γ point. They surprisingly found that the band splitting at the Γ point was not closely related to the structural transition, which may be controlled by magnetic frustration [27]. A high-resolution laser-based ARPES on FeSe recently uncovered a highly anisotropic Fermi surface around the Γ point. This is possibly related to the low-temperature orthorhombic phase (i.e., nematic state), because the splitting of $3d_{xz}/3d_{yz}$ bands is possibly due to anisotropic electron hopping. Moreover, the observation of an extremely anisotropic superconducting gap with two-fold symmetry was also associated with nematicity. Fitting the gap scale against momentum suggests a possible combination of extended *s*-wave and *d*-wave gap symmetry [28,29]. It can be found that many characterizations, at least, confirmed that the superconducting energy gap of FeSe was of the anisotropic type.

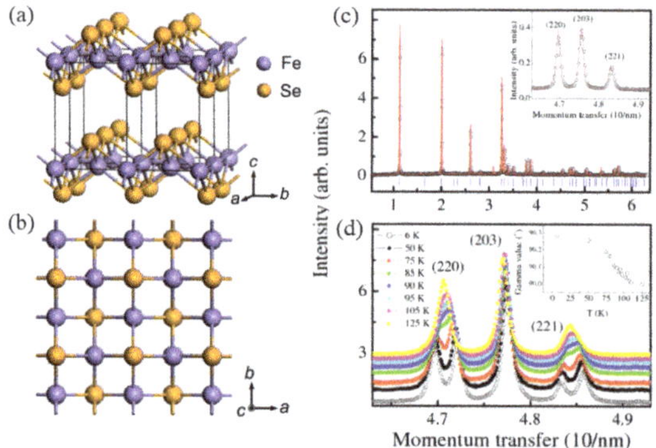

Figure 1. (**a**,**b**) Crystal structure of FeSe. (**c**) Observed (open black circle) and calculated (red solid line) powder diffraction intensities of FeSe$_{0.88}$ at 300 K using space group *P4/nmm*. (**d**) The temperature dependence of the γ-angle fitted with *P-1* symmetry. The single peak of the (2, 2, 0), (2, 0, 3), (2, 2, 1) reflection is seen in (**c**), splitting of the diffraction peaks into two peaks was observed (Figure reprinted from Hsu, F.C. et al. *Proc. Natl. Acad. Sci. U. S. A.* **2008**, 105, 14262–14264. Copyright 2008 by the National Academy of Sciences of the USA).

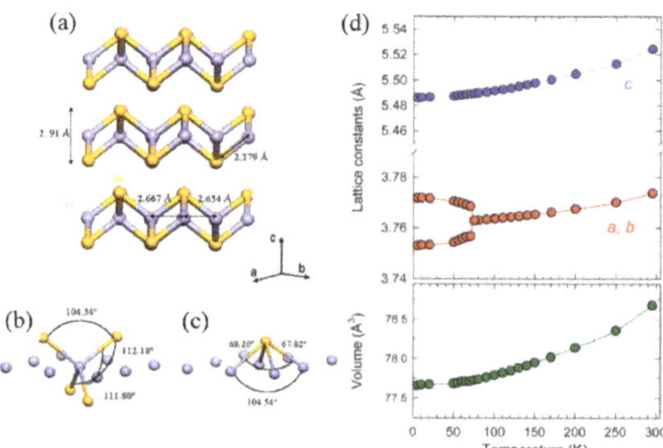

Figure 2. (**a**) Schematic diagram of the low-temperature orthorhombic structure of FeSe$_{0.92}$. (**b**,**c**) Geometry of the FeSe$_4$ tetrahedra and the SeFe$_4$ pyramids with three distinct Se–Fe–Se and Fe–Se–Fe bond angles, respectively. (**d**) Temperature dependent lattice constants of FeSe. (Figure reprinted from Margadonna, S. et al. *Chem. Commun. (Cambridge, U. K.).* **2008**, 5607–5609. Copyright 2008 by the Royal Society of Chemistry).

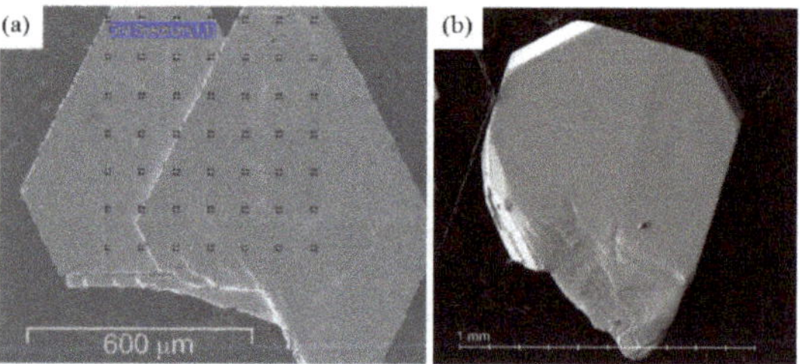

Figure 3. (**a**) Scanning electron micrographs of the as-grown FeSe$_{0.9}$ single crystal 2–3 mm in width and 50–60 µm in thickness by the high temperature flux-method. (**b**) The FeSe$_{0.96}$S$_{0.04}$ single crystal was obtained by chemical vapor deposition (Figure 3a reprinted from Vedeneev S.I. et al. *Phys. Rev. B* **2013**, 87, 134512. Copyright 2013 by American Physical Society. Figure 3b reprinted from Chareev, D. et al. *Cryst. Eng. Comm.* **2013**, 15, 1989–1993. Copyright 2013 by the Royal Society of Chemistry).

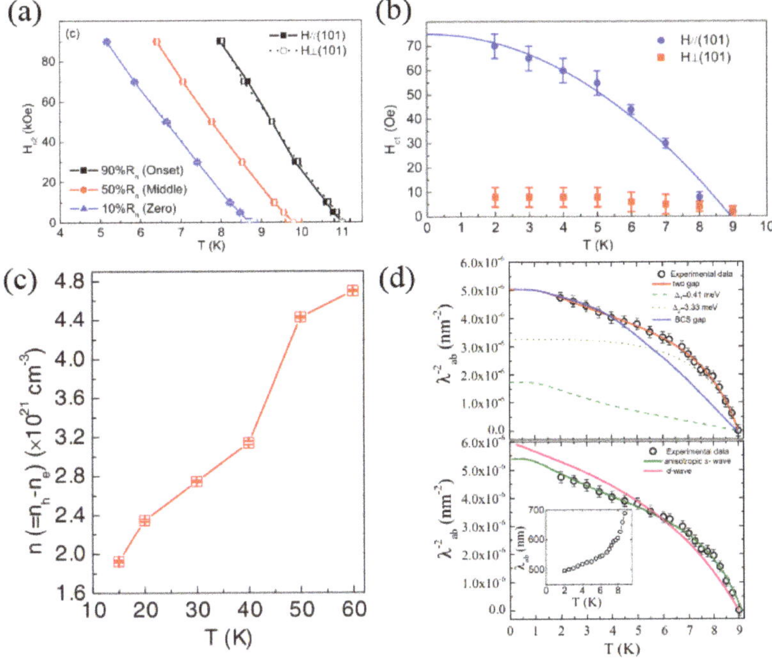

Figure 4. (a) Determination of the upper critical field $H_{c2}(0)$ from the temperature dependence of the resistivity $\rho(T)$ of β-FeSe single crystals for H//(101) and H⊥(101) directions. (b) Temperature-dependence of H_{c1} for both H//(101) and H⊥(101) directions. The solid blue line was the fitting curve using $H_{c1}(T) = H_{c1}(0)[1 - (T/T_c)^2]$ for H//(101). (c) Temperature dependence of carrier density n ($n_h - n_e$) of the β-FeSe crystal. (d) Fitted curve of two-gap model with anisotropic s-wave and d-wave models (Figure 4a,b reprinted from Lei H.C. et al. *Phys. Rev. B* **2011**, *84*, 014520. Copyright 2011 by American Physical Society. Figure 4c reprinted from Lei H.C. et al. *Phys. Rev. B* **2012**, 85, 094515. Copyright 2012 by American Physical Society. Figure 4d reprinted from Hafiez, M.A. et al. *Phys. Rev. B* **2013**, 88, 174512. Copyright 2013 by American Physical Society].

2.2. FeSe under Pressure and Chemical Substitution

The application of high pressure on FeSe exhibits a surprising effect on tuning SC and magnetism. Mizuguchi et al. first enhanced the T_c of polycrystalline FeSe to 13.5 K under 1.48 GPa with a positive pressure-coefficient of 9.1 K/GPa [30]. Linear-extrapolated $H_{c2}(0)$ was increased from 37 T to 72 T, implying the unconventional feature of SC. Millican et al. applied 0.6 GPa hydrostatic pressure to FeSe, and the structural parameters deduced from the high-resolution NPD pattern suggested that the phase transition was suppressed under pressure. A small bulk modulus, 31 GPa, suggested that FeSe is a soft material that can be effectively pressurized [31]. Garbarino et al. observed a new orthorhombic phase at a higher T_c of 34 K under 22 GPa [32]. Imai et al. performed the ^{77}Se nuclear magnetic resonance (NMR) measurement at 2.2 GPa and ascribed the strongly-enhanced AFM spin fluctuations to be the origin of the enhancement of T_c [33]. This picture was supported by a zero-field muon spin rotation (μSR) measurement, from which the SC and AFM order were simultaneously stabilized in the low pressure range (see Figure 5a) [34]. Meanwhile, a very small magnetic moment, 0.2 μ_B, was obtained at 2.4 GPa. In an extended pressure range, a clear dome-shaped superconducting phase diagram as a function of pressure was established, where the maximal T_c of 36.7 K showed up at 8.9 GPa. Above 12 GPa, the tetragonal FeSe changed into hexagonal FeSe (NiAs-type), which is a non-superconducting phase (Figure 5b) [35]. The in situ Mössbauer spectrum under pressure revealed that there was no long-range magnetic ordering in the whole pressure range. This observation is in stark contrast to the results of the NMR and μSR experiments.

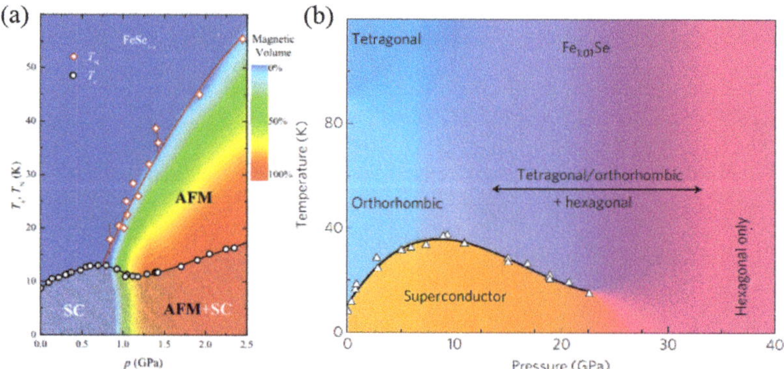

Figure 5. Electronic phase diagram of FeSe$_{1-x}$ under pressure. (**a**) Pressure dependence of the T_c, T_N, and the superconducting and magnetic volume fractions of FeSe$_{1-x}$. (**b**) Electronic phase diagram of Fe$_{1.01}$Se under a pressure range from 0 to 40 GPa. The maximum T_c observed was 36.7 K at 8.9 GPa (Figure 5a reprinted from Bendele M. et al. *Phys. Rev. B* **2012**, 85, 064517. Copyright 2012 by American Physical Society. Figure 5b reprinted from Medvedev S. et al. *Nat. Mater.* **2009**, 8, 630–633. Copyright 2009 by Macmillan Publishers Limited).

To clarify this, single crystals of FeSe have been used in in-situ high-pressure experiments. In 2015, Terashima et al. applied the high pressure of 2.72 GPa on FeSe single crystals and measured the resistance and ac magnetic susceptibility [36]. They confirmed that T_c generally increased under pressure, but dropped to a minimum T_c at ~1.2 GPa, which was consistent with the pressure-induced antiferromagnetic phase transition [37]. At pressures above 5 GPa, Sun et al. observed a sudden enhancement of T_c and suppression of T_N, indicating competition between SC and magnetic order (see Figure 6). Furthermore, an interesting linear-T resistivity was observed at higher pressure, which can be considered to be a signature of a quantum critical point [38]. Application of higher pressure would destabilize the AFM state and thus enhance the spin fluctuations, to some extent, which is similar to the mechanism enhancing T_c in FeAs-based superconductors. At the optimum T_c, the electrical transport behaviors are dominated by holes, which are enhanced near optimal pressure, implying a Fermi surface reconstruction due to AF ordering [39].

According to the reports by Yeh et al. Te-substitution of Se sites resulted in lattice expansion because of the larger ionic radius of Te, as shown in Figure 7 [40,41]. It also can enhance onset of superconductivity (T_c^{onset}) to 15.2 K and $H_{c2}(0)$ to 28.8 T as half of the Se was substituted. Later, a high-quality single crystal of Fe$_{1+y}$Se$_{1-x}$Te$_x$ was grown from the melt [42]. It showed that the single-phase can only exist in the Te-rich side. Bulk SC at 14 K in Fe$_{1+y}$Se$_{0.5}$Te$_{0.5}$ was confirmed simultaneously by resistivity, magnetic susceptibility, and heat capacity. Post-annealing on Te-rich samples could improve the filamentary SC into bulk SC [43–46]. Meanwhile, the excess Fe at the interstitial site not only suppresses SC, but also induces weakly-localized states. From the neutron pair density function analysis, Se and Te do not actually occupy the same crystallographic site, leading to a local symmetry around Fe that is lower than the average crystal symmetry (P4/nmm) in Fe$_{1+y}$Se$_{1-x}$Te$_x$ solid-solutions [47]. A comprehensive phase diagram of the SC and the structure of Fe$_{1+y}$Se$_{1-x}$Te$_x$ were constructed by Liu et al. where bulk SC emerged in the range of 50% to 70% Te doping [48]. The in-plane magnetic wave vector in FeTe for the AFM order is (π, 0); as the content of Se increases, the magnetic fluctuations gradually evolve into the (π, π) type associated with the scenario of Fermi surface nesting. This finding first revealed that both FeSe- and FeAs-based superconductors may share a similar mechanism for SC, assuming that the spin fluctuation picture is applicable.

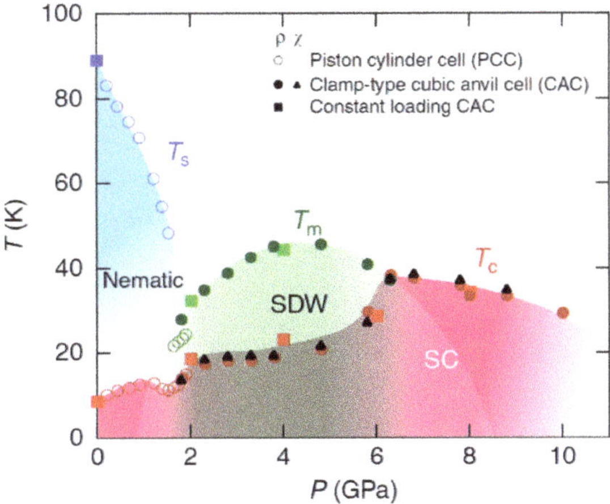

Figure 6. Superconducting phase diagram of bulk FeSe under pressure (Figure reprinted from Sun, J.P. et al. *Nat. Commun.* **2016**, *7*, 12146. Copyright 2016 by Macmillan Publishers Limited).

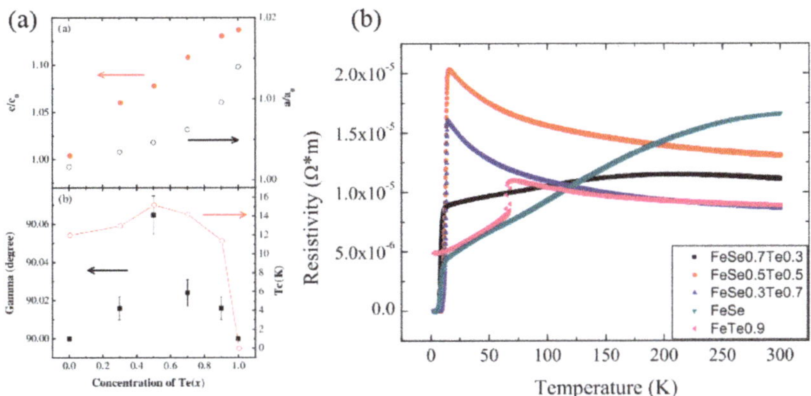

Figure 7. (**a**) Te-doping dependence of structural evolution and T_c of $FeSe_{1-x}Te_x$. (**b**) Temperature dependence of $FeSe_{1-x}Te_x$ resistance under zero magnetism (Figure reprinted from Yeh, K.W. et al. *J. Phys. Soc. Jpn.* **2008**, *77*, 19–22. Copyright 2008 by JPS).

Regarding S-doped FeSe, the soluble limit of S is smaller than 20%, according to a report from Mizuguchi et al. [49] Additionally, Abdel-Hafiez et al. studied the superconducting properties of high-quality single crystals of $FeSe_{1-x}S_x$ ($x = 0$, 0.04, 0.09, and 0.11). As the S concentration increases, the T_c determined from the onset of the diamagnetic signal increases from ~8.5 K to 10.7 K at $x = 0.11$. The normalized specific heat jump for $FeSe_{1-x}S_x$ was significantly larger than the limit (1.43) of the BCS model, which is a signature of strong-coupling SC [50]. By using ARPES, Watson et al. observed a smaller splitting of the $3d_{xz}$ and $3d_{yz}$ bands and a weaker anisotropy of the Fermi surface in the $FeSe_{1-x}S_x$ [51]. This means that isovalent S-substitution reduces the orbital ordering, correspondingly suppressing the structural transition temperature from 87 K to 58 K. Matsuura et al. investigated the pressure-dependent phase diagram of $FeSe_{1-x}S_x$, as shown in Figure 8 [52]. As S-content increased, the structural transition was quickly suppressed. Simultaneously, the pressure-induced SDW phase zone was gradually narrowed and the maximal T_c in each composition also decreased. This becomes a single dome-like SC in $FeSe_{0.83}S_{0.17}$, where the highest T_c was 30 K.

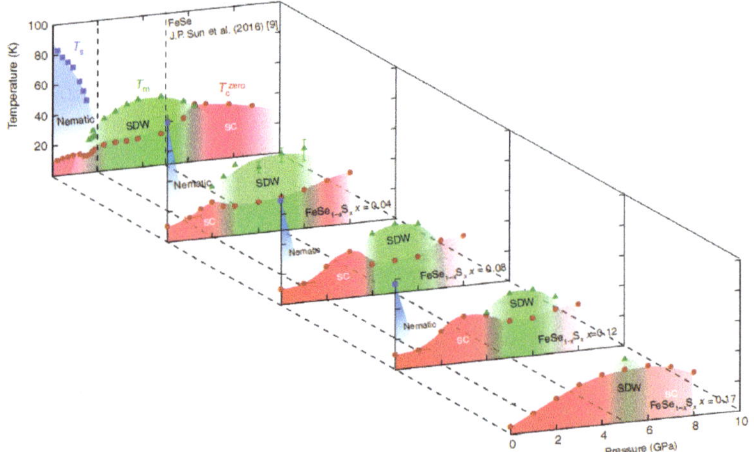

Figure 8. Temperature-pressure phase diagrams of FeSe$_{1-x}$S$_x$ (Figure reprinted from Matsuura, K. et al. *Nat. Commun.* **2017**, *8*, 1143. Copyright 2017 by Macmillan Publishers Limited).

3. Intercalated FeSe-Based Superconductors

In 2010, we discovered the alkali-intercalated FeSe-based superconductor with T_c ~30 K for the first time [53,54]. Through Rietveld refinements against PXRD, an average structure of K$_{0.8}$Fe$_2$Se$_2$ (i.e., 122-type), was determined, which was iso-structural, to be known BaFe$_2$As$_2$, as seen in Figure 9. From the temperature dependence of resistance and magnetization, the T_c ^{onset} was identified as 30.1 K. The dominant carriers were n-type, suggesting an electron-doped superconductor. This was the first report of enhancing T_c in an FeSe-based superconductor up to 30 K without applying high pressure. Following this discovery, several groups quickly identified many 122-type FeSe-based superconductors such as Cs$_{0.8}$(FeSe$_{0.98}$)$_2$ (T_c ~27 K) [55,56], Rb$_{0.88}$Fe$_{1.81}$Se$_2$ (T_c = 32 K) [57], Tl$_{0.58}$Rb$_{0.42}$Fe$_{1.72}$Se$_2$ (T_c ~32 K) [58], and (Tl, K)Fe$_x$Se$_2$ (T_c ~31 K) [59]. A new family of materials thereby came into the awareness of the superconducting community.

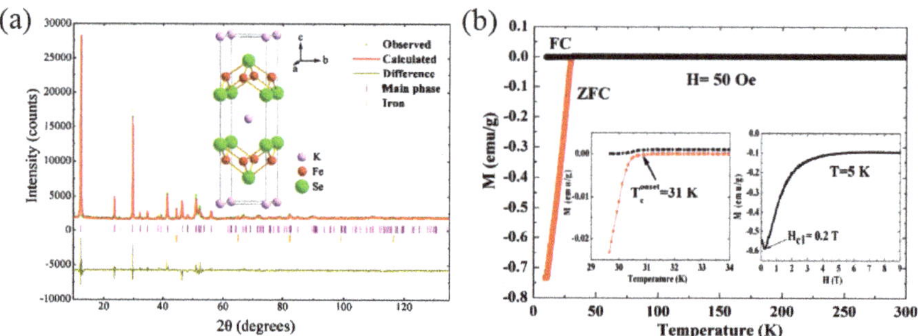

Figure 9. (**a**) PXRD pattern and Rietveld refinements profile of K$_{0.8}$Fe$_2$Se$_2$. Crystal structure is shown as inset. (**b**) Temperature dependence of magnetization of the K$_{0.8}$Fe$_2$Se$_2$ single crystal (Figure reprinted from Guo J.G. et al. *Phys. Rev. B* **2010**, *82*, 180520. Copyright 2010 by American Physical Society).

Transmission electron microscopy (TEM) on K$_{0.8}$Fe$_x$Se$_2$ and KFe$_x$Se$_2$ ($1.7 \leq x \leq 1.8$) samples revealed that complex phase-separation emerge in such AFe$_2$Se$_2$ superconductors. The superconducting phase is likely to be K$_x$Fe$_2$Se$_2$, which was later confirmed by STM results. The Fe-vacancy ordered phase K$_2$Fe$_4$Se$_5$ showed a $\sqrt{5} \times \sqrt{5} \times 1$ superstructure [60]. The NPD on K$_x$Fe$_2$Se$_2$ samples revealed that the superstructure phases A$_2$Fe$_4$Se$_5$ existed in all K, Cs, Rb, (Tl, Rb), and (Tl, K) intercalated FeSe-based superconductors [61]. Li et al. proposed that the existence of Fe vacancies could locally destroy SC [62], and suggested that the $K_2Fe_4Se_5$ layer may be indispensable to regulate superconductivity in KFe_2Se_2 by providing charge carriers [63]. More superstructures associated with different Fe-vacancy and alkali-metal-ordering were later revealed later [64–67].

Phase separation hinders the investigation of the intrinsic property of FeSe-based 122-type superconductors. A low-temperature route, named the liquid-ammonia method, was proposed to synthesize phase-pure superconductors. A series of alkali and ammonia co-intercalated compounds were synthesized such as $Li(NH_3)_yFe_2Se_2$, $Na(NH_3)_yFe_2Se_2$, $Ca_{0.5}(NH_3)_yFe_2Se_2$, $Sr_{0.8}(NH_3)_yFe_2Se_2$, $Ba_{0.8}(NH_3)_yFe_2Se_2$, $Eu(NH_3)_yFe_2Se_2$, and $Yb(NH_3)_yFe_2Se_2$, with a T_c of 30–46 K [68–70]. Taking $Na(NH_3)_yFe_2Se_2$ as an example, the crystal structure is composed of edge-sharing $FeSe_4$-tetrahedra layers and Na atoms between the $FeSe_4$ layers, and the T_c ^{onset} is 46 K, as shown in Figure 10a,b. In addition, these authors built a discrete superconducting phase diagram against K content, as seen in Figure 10c,d, where the low intercalated content could enhance the T_c to 44 K. After increasing the K content to 0.6, the T_c was lowered to 30 K [71]. This trend is well reproduced by the density function theory and Monte Carlo simulations [72].

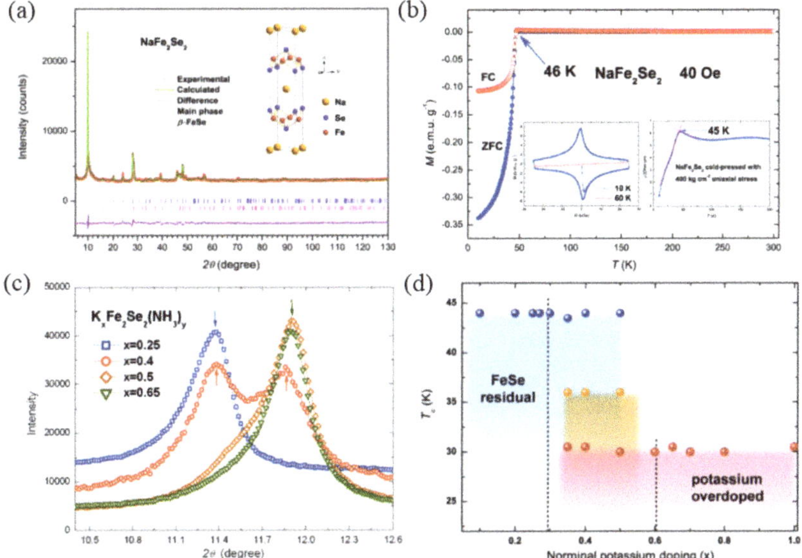

Figure 10. (**a**) PXRD pattern and Rietveld refinements profile of $NaFe_2Se_2$ at ambient temperature. Inset shows the crystal structure of $NaFe_2Se_2$. (**b**) Temperature dependence of magnetization of $NaFe_2Se_2$ polycrystalline sample. (**Left**) inset shows superconducting loops of $NaFe_2Se_2$ at 10 K and 60 K, respectively. (**Right**) inset shows temperature dependence of the electrical resistance of cold-pressed $NaFe_2Se_2$. (**c**) Enlargement of (002) peak of PXRD patterns for $K_xFe_2Se_2(NH_3)_y$ ($x = 0.25$, 0.4, 0.5, and 0.65) samples measured at 297 K. (**d**) T_cs of $K_xFe_2Se_2(NH_3)_y$ as a function of nominal potassium content (Figure 10a,b reprinted from Ying, T.P. et al. *Sci. Rep.* **2012**, 2, 426. Copyright 2012 by Macmillan Publishers Limited. Figure 10c,d reprinted from Ying, T. P. et al. *J. Am. Chem. Soc.* **2013**, 135, 2951–2954. Copyright 2013 by American Chemical Society).

Later, the high-resolution NPD technique was adopted to determine the crystal structure of $Li_x(NH_3)_yFeSe$, where the Li and NH_3 are located in between the FeSe layer [73]. Jin et al. also determined the structure of $Na_{0.39}(C_2N_2H_8)_{0.77}Fe_2Se_2$ in this way, as shown in Figure 11 [74]. The existence of organic molecules seems not to be related to SC, but can drastically alter the geometry of FeSe, even changing the structural symmetry due to rotations of the long-chain of the C atom [75].

Lu et al. synthesized an intercalated superconductor $(Li_{0.8}Fe_{0.2})OHFeSe$ by the hydrothermal method. The T_c was ~41 K and the precise structure parameters were determined by x-ray and NPD. the details of which are shown in Figure 12 [76]. The T_c of different FeSe intercalates imply that T_c is proportional to the interlayer spacing below 9.0 Å. Once the spacing is above 9.0 Å, T_c seems to be independent on this spacing [77]. Therefore, the doped electron count is more important for determining T_c. Thus, studies of metal ions and molecular co-intercalated FeSe-based compounds are important to enhance T_c and understand the effect of structural change on SC [78,79]. Additionally, under high pressure, a second superconducting phase with higher T_c was observed in $K_xFe_2Se_2$, which can be ascribed to the quantum critical phase transition [80,81]. Such pressure-induced SC was also verified in $(Li_{0.8}Fe_{0.2})OHFeSe$ [82] and in $Li_x(NH_3)_yFe_2Se_2$ [83].

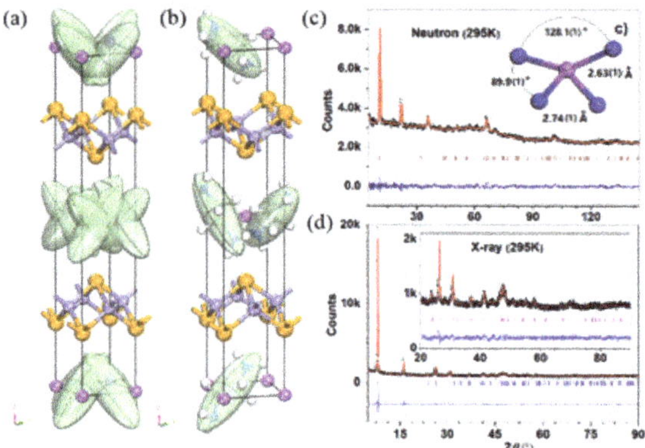

Figure 11. (**a–b**) Crystal structure of $Na_{0.39}(C_2N_2H_8)_{0.77}Fe_2Se_2$. (**c**) Rietveld refinements of NPD pattern. Inset is the atomic geometry of $FeSe_4$ tetrahedra. (**d**) Rietveld refinements of PXRD pattern. Inset shows the expanded range of 20°–90°. [Figure reprinted from Jin S.F. et al. *Chem. Commun. (Cambridge, U. K.)* **2017**, 53, 9729-9732. Copyright 2017 by The Royal Society of Chemistry.].

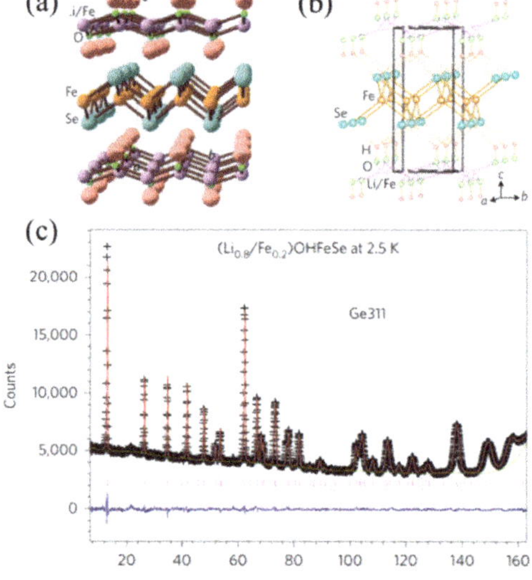

Figure 12. (**a,b**) Crystal structure of $(Li_{0.8}Fe_{0.2})OHFeSe$. (**c**) Rietveld refinements of PND pattern of $(Li_{0.8}Fe_{0.2})OHFeSe$ (Figure reprinted from Lu X.F. et al. *Nat. Mater.* **2015**, 14, 325–329. Copyright 2015 by Macmillan Publishers Limited).

Sizeable single-crystals of $A_xFe_2Se_2$ and $(Li_{0.8}Fe_{0.2})OHFeSe$ superconductors can be easily grown by the self-flux and hydrothermal method. Thus, the important Fermi surface topology can be detected by ARPES experiments. The most important finding is that the hole-pocket at the Γ point sinks tens of meV below the Fermi level and the unique Fermi surface only contains electronic pockets [84]. In terms of the approximation of a rigid band shift, it is a consequence of electron-doping, which lifts up the Fermi energy. This topology of the Fermi surface has also been observed in $(Li_{0.8}Fe_{0.2})OHFeSe$ (see Figure 13) [85,86]. Such a Fermi surface challenges the established picture of electron-hole nesting for inducing SC in FeAs-based superconductors [9,10,87]. At the same time, the ARPES measurements confirmed that the superconducting gap was of the isotropic type, with a magnitude of 10.3 meV. This nodeless gap suggests that a conventional *s*-wave pairing could better describe the origin of SC [88].

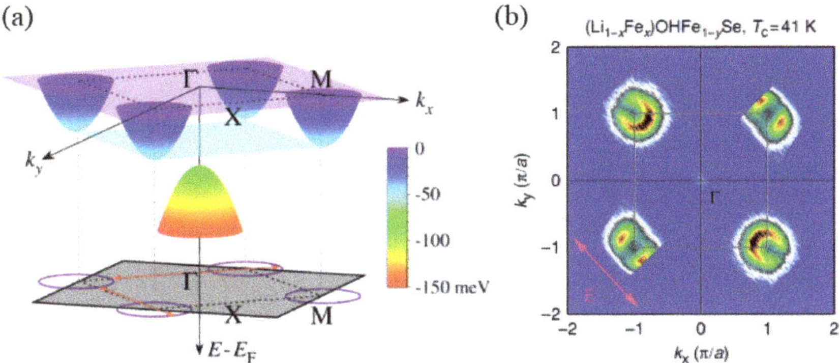

Figure 13. (a) Electronic band structure of $K_{0.8}Fe_{1.7}Se_2$ and possible (π, π) scattering vector. (b) Fermi surface of $(Li_{0.84}Fe_{0.16})OHFe_{0.98}Se$ measured at 20 K (Figure reprinted from Qian T. et al. *Phys. Rev. Lett.* **2011**, 106, 187001. Copyright 2011 by American Physical Society. Reprinted from Zhao L. et al. *Nat. Commun.* **2016**, 7, 10608. Copyright 2016 by Macmillan Publishers Limited).

4. Superconductivity of FeSe Film

In 2012, Wang et al. reported that a single unit cell (SUC) FeSe film grown on a $SrTiO_3$ (STO) substrate by molecular beam epitaxy (MBE) exhibited a very large superconducting gap (Δ) of 20 meV. Assuming that the superconducting mechanism of both FeSe film and bulk FeSe is the same, the estimated T_c of the SUC FeSe film may be larger than 77 K [89]. The SUC FeSe film had an atomically smooth surface after Se-flux treatment. The temperature-dependent resistance showed a little lower T_c^{onset} of 53 K as plotted in Figure 14a. The upper inset in Figure 14a shows that the SC is slowly suppressed by the external magnetic field. The tunneling spectrum in the SUC FeSe film clearly exhibited a ~20 meV superconducting gap, as seen in Figure 14b. However, the double unit cell (DUC) FeSe film did not exhibit SC at all (see Figure 14c). It seems that the interfacial effect offers an effective way to realize high T_c in FeSe. Ge et al. reported a rather high T_c, above 100 K, in FeSe/STO film by using in situ four-point probe electrical transport measurements. This T_c is the highest value in all reported Fe-based superconductors, which is a rather exciting result that deserves further investigation [90]. Subsequently, many FeSe thin films grown on different substrates have been studied in order to understand the mechanism of enhancing SC.

He et al. studied the topology of the Fermi surface of the SUC FeSe film grown on a STO substrate by changing the carrier concentration through a different annealing procedure [91]. Surprisingly, they found that two competitive phases, a non-superconducting phase at low-doping level and another superconducting phase at high-doping level, appeared during the annealing process. The electronic structure and the superconducting gap were tuned in a larger range by adjusting the annealing time and temperature. As depicted in Figure 15a–d, it was found that the magnitude of the superconducting gap increased from ~10 meV to ~19 meV. Meanwhile, the relation between the superconducting gap size and temperature could be well described by the BCS theory, which is displayed as a green line in Figure 15e–h). Meanwhile, Tan et al. also grew a high-quality thin FeSe film on a STO substrate and measured the Fermi surface and band dispersion [92]. They found that a short-range SDW appeared in the thin FeSe film (50 unit-cell). As depicted in Figure 16a, the SC arises while the spin density wave is suppressed. This phase diagram reveals that there is a unified trend in SUC FeSe/STO and other Fe-based superconductors by controlling the doping level and lattice constants. It also implies that a higher T_c will appear in the FeSe film if the lattice constants are expanded and higher electronic doping is introduced. Rebec et al. compared the T_c and the scale of superconducting gap in 60 unit-cell FeSe/STO (001), SUC FeSe/STO (001), and 3 UC FeSe/STO (001) coated by potassium atoms and SUC FeSe/TiO$_2$ (100). By analyzing the ARPES results, they found that electron doping plays an important role in inducing high T_c [93].

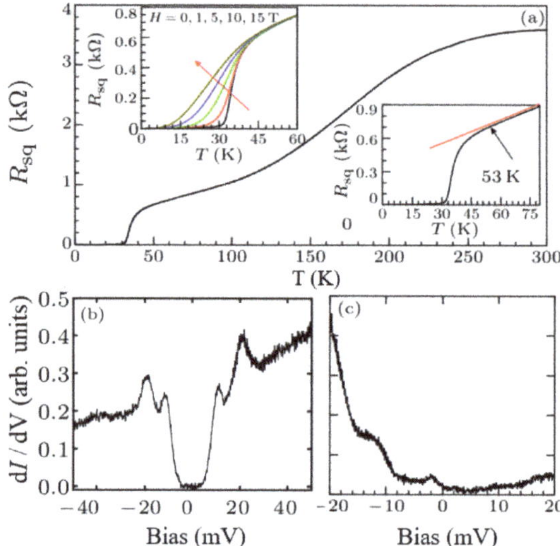

Figure 14. (**a**) Temperature dependence of square resistivity of a 5-UC-thick FeSe film on insulating STO (001). (**b**) Tunneling spectrum measurement of the 1-UC-thick FeSe film on STO (001) at 4.2 K. (**c**) dI/dV spectrum of the 2-UC-thick FeSe film on insulating STO (001) at 4.2 K. (Figure reprinted from Wang, Q.Y. et al. *Chin. Phys. Lett.* **2012**, *29*, 037402. Copyright 2012 by Chinese Physical Society and IOP Publishing Ltd.).

There is an interaction between electrons from FeSe and the oxygen optical phonons of the STO substrate, thus enhancing the electron-phonon coupling [94]. Figure 17 suggests that there is a close relationship between interfacial electron–phonon coupling and T_c enhancement. Both experimental and calculation results revealed that interfacial electron–phonon coupling could account for the enhancement of T_c in the SUC FeSe/STO film. To find out the main factor in enhancing SC, Ding et al. grew SUC FeSe films on a TiO$_2$ (001) substrate [95]. As depicted in Figure 18, a TiO$_2$ film with a thickness of 15 nm was grown on a STO substrate. On the TiO$_2$ surface, a 4 × 1 reconstruction of the oxygen vacancies (white spots) can clearly be seen. The density of the oxygen vacancies in the as-grown TiO$_2$ surface was about 4.6×10^{-2} per nm^2, and post-annealing reduced the vacancies of TiO$_2$ to 6.1×10^{-3} per nm^2, as shown in Figure 18a–c. The SUC FeSe film on TiO$_2$ also exhibited a clear superconducting feature in the dI/dV spectrum, but the double unit cell (DUC) FeSe was not superconducting (see Figure 18d). In the SUC FeSe sample grown on a TiO$_2$ substrate with different oxygen vacancies, there was no change in the superconducting gap (see Figure 18e). Thus, the density of interfacial oxygen vacancies has a limited effect on enhancing the interfacial charge transfer and T_c.

Zhou et al. also grew SUC FeSe films on a MgO (001) substrate, and the onset T_c was 18 K, as inferred by transport measurement. [96] The atomic image illustrates that the SUC FeSe film grown on the MgO (001) substrate was along FeSe (001). A 1.3% in-plane tensile strain was induced by the lattice mismatch of the FeSe and the MgO (001) substrate. Scanning transmission electron microscopy (STEM) images of the interfacial structure demonstrated that Mg atoms could be replaced by Fe atoms, and the result of the calculations showed that the MgO film near the interface becomes electron doped, which shows the capability of enhancing the charge-transfer effect.

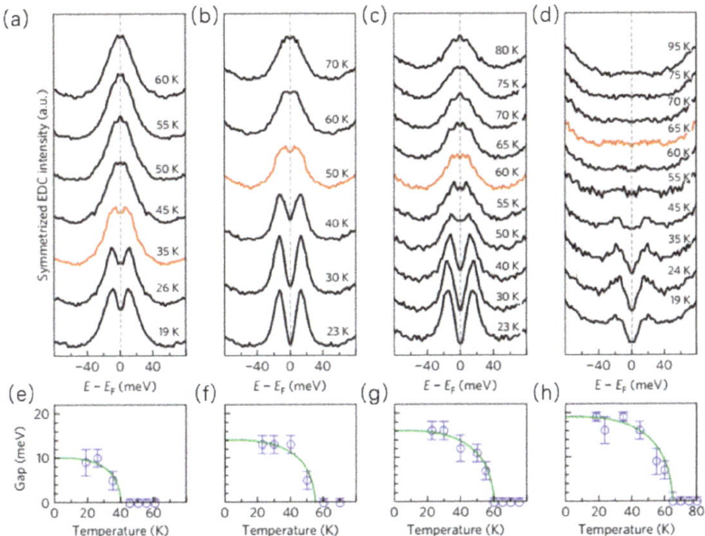

Figure 15. (**a–d**) Symmetrized energy distribution curves (EDCs) of the Fermi surface near the M point in the Brillouin zone with various annealing process. (**e–h**) Temperature dependence of the superconducting gap. Green lines are fitting curves based on the BCS model (Figure reprinted from He, S.L. et al. *Nat. Mater.* **2013**, 12, 605–610. Copyright 2013 by Macmillan Publishers Limited.).

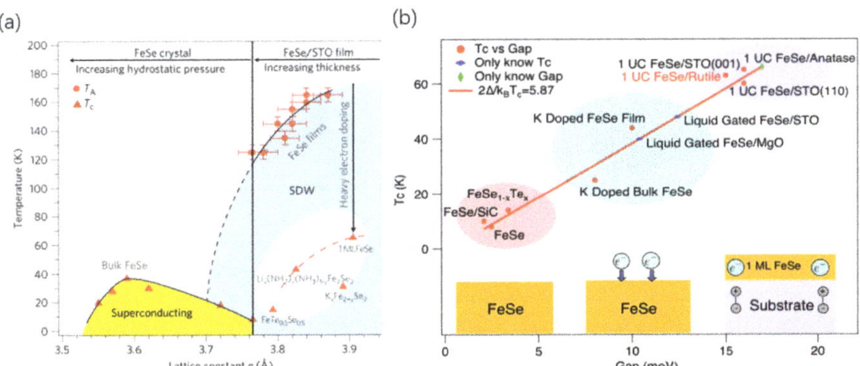

Figure 16. (**a**) The proposed phase diagram of various FeSe film and bulk FeSe. (**b**) T_c and superconducting gap of bulk FeSe, FeSe grown on the SiC (0001) substrate, K-doped thin FeSe film, K-doped FeSe film grown on the STO (001) substrate, single unit cell (SUC) FeSe/STO (001), and FeSe/STO (110). (Figure 16a reprinted from Tan, S.Y. et al. *Nat. Mater.* **2013,** 12, 634–640. Copyright 2013 by Macmillan Publishers Limited. Figure 16b reprinted from Rebec, S.N. et al. Phys. Rev. Lett. 2017, 118, 067002. Copyright 2017 by American Physical Society).

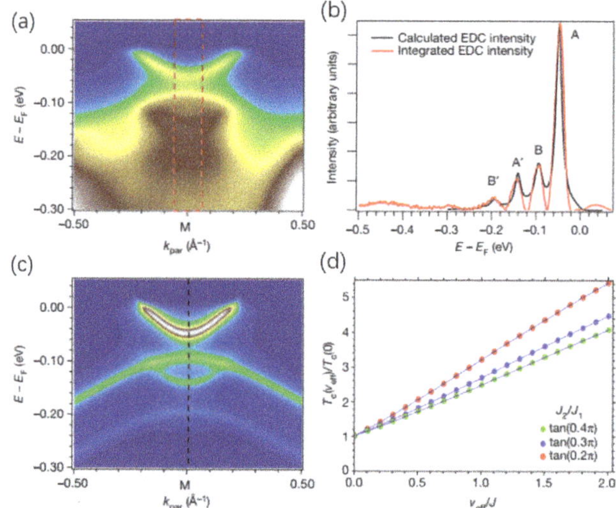

Figure 17. Electron-phonon coupling effect and measurement of T_c enhancement. (**a**) High-statistics scan at the M point of the first Brillouin zone measured at 10 K. (**b**) The EDC after background subtraction and the calculated counterpart. (**c**) Theoretical calculation involving hole and electron bands coupled to a dispersion phonon mode via model spectral functions, and the dashed line is the symmetrized energy distribution curves. (**d**) T_c enhancement as a function of electron–phonon coupling strength (V_{eff}/J). (Figure 17 reprinted from Lee, J.J. et al. *Nature* **2014**, 515, 245–248. Copyright 2014 by Macmillan Publishers Limited).

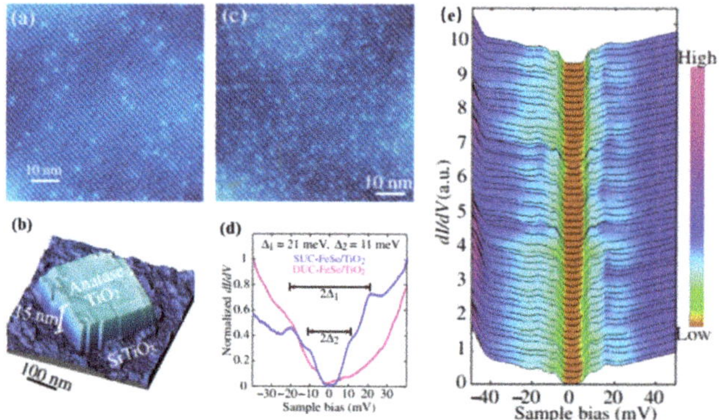

Figure 18. (**a**) STM topography of TiO_2 with lower density of oxygen vacancies after annealing (bright spots). (**b**) STM topography in a TiO_2 island supported by the STO substrate. (**c**) Enlarged STM topography acquired on a TiO_2 island with oxygen vacancies (without annealing). (**d**) Low energy dI/dV spectra taken on the SUC and DUC FeSe films. (**e**) A series of dI/dV spectra acquired from the FeSe surface along one direction (Figure 18 reprinted from Ding, H. et al. *Phys. Rev. Lett.* **2016**, 117. Copyright 2016 by American Physical Society).

Miyata et al. observed SC with a T_c of 48 K in a multilayer FeSe film by coating potassium onto its surface [97]. The electronic structure of the SUC FeSe films are shown in Figure 19. The Fermi surface of the SUC FeSe film consists of a large electron pocket at the M point of the Brillouin zone, which is consistent with those of $K_xFe_2Se_2$ and LiOHFeSe. However, the Fermi surface of the triple unit-cell FeSe film included a hole pocket at the Γ point and a tiny electron pocket near the M point of the

Brillouin zone (see Figure 19d–f), which is similar to that of bulk FeSe. After coating more K atoms, a downward shift of hole-like band occurred at the Γ point. The hole-pocket at the Γ point thereby gradually disappeared, as shown in Figure 19g–o. The total doped-electron count was about $0.11e^-$ in the triple unit-cell FeSe film, and a superconducting gap of 8 meV opened at T_c = 46 K and closed above 51 K. Obviously, electron doping by coating K atoms is a useful strategy to induce high T_c in multilayer FeSe films.

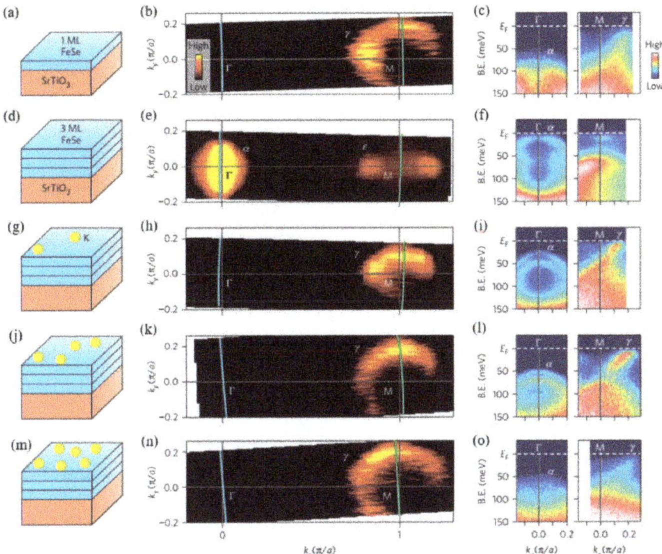

Figure 19. (**a,d,g,j,m**) Schematic diagram of the FeSe films with different thicknesses and content grown on the STO substrate. (**b,e,h,k,n**) Fermi surfaces mapped out by ARPES. Green and blue lines are momentum cuts around the Γ and M points. (**c,f,i,l,o**) ARPES intensity image near the Fermi energy (E_F) as a function of binding energy and wav-vector along the cuts close to Γ and M. (Figure 19 reprinted from Miyata, Y. et al. *Nat. Mater.* **2015**, 14, 775–779. Copyright 2015 by Macmillan Publishers Limited).

Interfacial electron–phonon coupling has also been proposed to explain the enhancement of high T_c in SUC FeSe/STO. However, the evidence that can directly connect interfacial electron–phonon coupling intensity (EPI) with high T_c is still not sufficiently clear. Song et al. studied thin films of FeSe by replacing ^{16}O with its isotope ^{18}O [98]. Surprisingly, they found that the energy difference between the electron-type band and its replica band was approximately in proportion to the inverse square root of the mass of oxygen atoms. Therefore, the SC is highly related to interfacial EPI. As depicted in Figure 20, superconducting gaps Δ_1 and Δ_2 were ~9.5 meV while the ratio η (i.e., the EPI) is close to zero, which is basically equal to the superconducting gap of the K-dosed multilayer FeSe film. Thus, it demonstrated that a synergistic effect between possible spin fluctuations and enhanced EPI results in such high T_c in FeSe/STO. It is worth noting that theorists could qualitatively describe the Fermi surface, hidden magnetic order, replica bands, and the superconductivity isotropic Cooper pairs in heavily electron-doped FeSe by an extended Hubbard model [99,100]. Such explanations could deepen the understanding of the superconducting mechanism and may help us explore more high T_c superconductors.

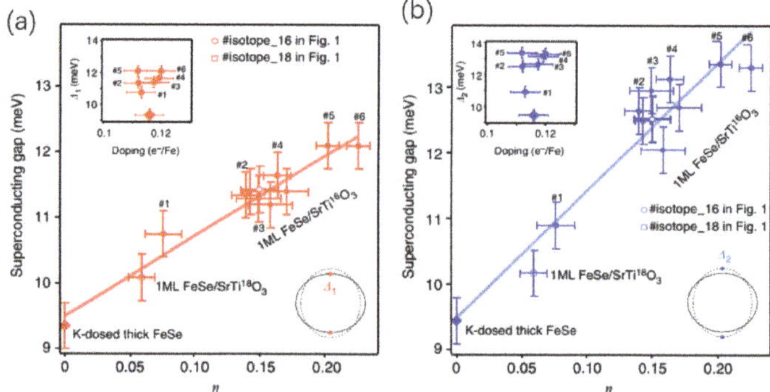

Figure 20. (**a**,**b**) Relationship between the superconducting gap Δ_1, Δ_2 and the intensity ratio (η) between the side band and the main band. The insets show the connection between the superconducting gap sizes and doping for corresponding samples. (Figure 20 reprinted from Song, Q. et al. *Nat. Commun.* **2019**, *10*, 758. Copyright 2019 by Macmillan Publishers Limited).

5. Conclusions

The finding of a series of surprisingly high T_c in FeSe-based materials is a significant breakthrough in the superconducting community and sets up a very interesting superconducting family. However, the microscopic mechanism of the enhancement of T_c through multiple treatments is still elusive. Many kinds of effects have been proposed to explain the enhancement in T_c such as coupling between the electrons and phonons of the substrate, the tensile strain effect introduced by lattice mismatch between FeSe film and the substrate, and the charge transfer, *etc*. More subtle experiments should be conducted to figure out which effect plays the decisive role. In addition, as FeSe thin films are very thin and are easily oxidized, the transport and magnetic properties are hard to measure. Therefore, new methods and more novel experimental tools should be developed to detect the intrinsic physical properties.

Author Contributions: Investigation, K.Z., J.W.; Y.S. writing—original draft preparation, K.Z.; J.W.; L.G.; J.-g.G.; writing—review and editing, K.Z.; J.W.; J.-g.G.; funding acquisition, J.-g.G.

Funding: This research was funded by the National Natural Science Foundation of China, grant number 51922105 and 51772322, the National Key Research and Development Program of China, grant number 2016YFA0300600 and 2017YFA0304700, and the Chinese Academy of Sciences, grant number QYZDJ-SSW-SLH013.

Conflicts of Interest: The authors declare no conflict of interest.

References

1. Kamihara, Y.; Watanabe, T.; Hirano, M.; Hosono, H. Iron-Based Layered Superconductor La[$O_{1-x}F_x$]FeAs (x = 0.05-0.12) with T_c = 26 K. *J. Am. Chem. Soc.* **2008**, *130*, 3296. [CrossRef]

2. Takahashi, H.; Igawa, K.; Arii, K.; Kamihara, Y.; Hirano, M.; Hosono, H. Superconductivity at 43 K in an iron-based layered compound $LaO_{1-x}F_x$FeAs. *Nature* **2008**, *453*, 376–378. [CrossRef] [PubMed]

3. Chen, X.H.; Wu, T.; Wu, G.; Liu, R.H.; Chen, H.; Fang, D.F. Superconductivity at 43 K in $SmFeAsO_{1-x}F_x$. *Nature* **2008**, *453*, 761–762. [CrossRef] [PubMed]

4. Ren, Z.A.; Lu, W.; Yang, J.; Yi, W.; Shen, X.L.; Li, Z.C.; Che, G.C.; Dong, X.L.; Sun, L.L.; Zhou, F.; et al. Superconductivity at 55 K in iron-based F-Doped layered quaternary compound Sm[$O_{1-x}F_x$]FeAs. *Chin. Phys. Lett.* **2008**, *25*, 2215–2216. [CrossRef]

5. Hsu, F.C.; Luo, J.Y.; Yeh, K.W.; Chen, T.K.; Huang, T.W.; Wu, P.M.; Lee, Y.C.; Huang, Y.L.; Chu, Y.Y.; Yan, D.C.; et al. Superconductivity in the PbO-type structure α-FeSe. *Proc. Natl. Acad. Sci. USA.* **2008**, *105*, 14262–14264. [CrossRef]

6. Kasahara, S.; Shibauchi, T.; Hashimoto, K.; Ikada, K.; Tonegawa, S.; Okazaki, R.; Shishido, H.; Ikeda, H.; Takeya, H.; Hirata, K.; et al. Evolution from non-Fermi- to Fermi-liquid transport via isovalent doping in $BaFe_2(As_{1-x}P_x)_2$ superconductors. *Phys. Rev. B* **2010**, *81*, 184519. [CrossRef]

7. Arsenijević, S.; Hodovanets, H.; Gaál, R.; Forró, L.; Bud'ko, S.L.; Canfield, P.C. Signatures of quantum criticality in the thermopower of $Ba(Fe_{1-x}Co_x)_2As_2$. *Phys. Rev. B* **2013**, *87*, 224508. [CrossRef]

8. De la Cruz, C.; Huang, Q.; Lynn, J.W.; Li, J.; Ratcliff, W.; Zarestky, J.L.; Mook, H.A.; Chen, G.F.; Luo, J.L.; Wang, N.L.; et al. Magnetic order close to superconductivity in the iron-based layered $LaO_{1-x}F_xFeAs$ systems. *Nature* **2008**, *453*, 899–902. [CrossRef]

9. Hirschfeld, P.J.; Korshunov, M.M.; Mazin, I.I. Gap symmetry and structure of Fe-based Superconductors. *Rep. Prog. Phys.* **2011**, *74*, 124508. [CrossRef]

10. Johnson, P.D.; Xu, G.Y.; Yin, W.G. *Iron-Based Superconductivity*; Series in Materials Science; Springer: Basel, Switzerland, 2015; pp. 211–255.

11. Margadonna, S.; Takabayashi, Y.; McDonald, M.T.; Kasperkiewicz, K.; Mizuguchi, Y.; Takano, Y.; Fitch, A.N.; Suard, E.; Prassides, K. Crystal structure of the new $FeSe_{1-x}$ superconductor. *Chem. Commun.* **2008**, 5607–5609. [CrossRef]

12. Chu, J.H.; Analytis, J.G.; Greve, K.D.; McMahon, P.L.; Islam, Z.; Yamamoto, Y.; Fisher, I.R. In-plane resistivity anisotropy in an underdoped iron arsenide superconductor. *Science* **2010**, *329*, 824. [CrossRef] [PubMed]

13. Wang, Q.; Shen, Y.; Pan, B.; Zhang, X.; Ikeuchi, K.; Iida, K.; Christianson, A.D.; Walker, H.C.; Adroja, D.T.; Abdel-Hafiez, M.; et al. Magnetic ground state of FeSe. *Nat. Commun.* **2016**, *7*, 12182. [CrossRef] [PubMed]

14. Pomjakushina, E.; Conder, K.; Pomjakushin, V.; Bendele, M.; Khasanov, R. Synthesis, crystal structure, and chemical stability of the superconductor $FeSe_{1-x}$. *Phys. Rev. B* **2009**, *80*, 024517. [CrossRef]

15. Williams, A.J.; McQueen, T.M.; Cava, R.J. The stoichiometry of FeSe. *Solid State Commun.* **2009**, *149*, 1507–1509. [CrossRef]

16. McQueen, T.M.; Williams, A.J.; Stephens, P.W.; Tao, J.; Zhu, Y.; Ksenofontov, V.; Casper, C.; Cava, R.J. Tetragonal-to-orthorhombic structural phase transition at 90 K in the superconductor $Fe_{1.01}Se$. *Phys. Rev. Lett.* **2009**, *103*, 057002. [CrossRef]

17. Chen, T.K.; Chang, C.C.; Chang, H.H.; Fang, A.H.; Wang, C.H.; Chao, W.H.; Tseng, C.M.; Lee, Y.C.; Wu, Y.R.; Wen, M.H.; et al. Fe-vacancy order and superconductivity in tetragonal β-$Fe_{1-x}Se$. *Proc. Natl. Acad. Sci. USA.* **2014**, *111*, 63–68. [CrossRef]

18. Keimer, B.; Kivelson, S.A.; Norman, M.R.; Uchida, S.; Zaanen, J. From quantum matter to high-temperature superconductivity in copper oxides. *Nature* **2015**, *518*, 179–186. [CrossRef]

19. Mok, B.H.; Rao, S.M.; Ling, M.C.; Wang, K.J.; Ke, C.T.; Wu, P.M.; Chen, C.L.; Hsu, F.C.; Huang, T.W.; Luo, J.Y.; et al. Growth and investigation of crystals of the new superconductor α-FeSe from KCl Solutions. *Cryst. Growth Des.* **2009**, *9*, 3260–3264. [CrossRef]

20. Chareev, D.; Osadchii, E.; Kuzmicheva, T.; Lin, J.Y.; Kuzmichev, S.; Volkova, O.; Vasiliev, A. Single crystal growth and characterization of tetragonal $FeSe_{1-x}$ superconductors. *Cryst. Eng. Comm.* **2013**, *15*, 1989–1993. [CrossRef]

21. Lei, H.C.; Hu, R.W.; Petrovic, C. Critical fields, thermally activated transport, and critical current density of β-FeSe single crystals. *Phys. Rev. B* **2011**, *84*, 014520. [CrossRef]

22. Lei, H.C.; Graf, D.; Hu, R.W.; Ryu, H.J.; Choi, E.S.; Tozer, S.W.; Petrovic, C. Multiband effects on β-FeSe single crystals. *Phys. Rev. B* **2012**, *85*, 094515. [CrossRef]

23. Lin, J.Y.; Hsieh, Y.S.; Chareev, D.A.; Vasiliev, A.N.; Parsons, Y.; Yang, H.D. Coexistence of isotropic and extended s-wave order parameters in FeSe as revealed by low-temperature specific heat. *Phys. Rev. B* **2011**, *84*, 220507. [CrossRef]

24. Hafiez, M.A.; Ge, J.; Vasiliev, A.N.; Chareev, D.A.; Vondel, J.V.; Moshchalkov, V.V.; Silhanek, A.V. Temperature dependence of lower critical field $H_{c1}(T)$ shows nodeless superconductivity in FeSe. *Phys. Rev. B* **2013**, *88*, 174512. [CrossRef]

25. Subedi, A.; Zhang, L.; Singh, D.J.; Du, M.H. Density functional study of FeS, FeSe, and FeTe: Electronic structure, magnetism, phonons, and superconductivity. *Phys. Rev. B* **2008**, *78*, 134514. [CrossRef]

26. Shimojima, T.; Suzuki, Y.; Sonobe, T.; Nakamura, A.; Sakano, M.; Omachi, J.; Yoshioka, K.; Gonokami, M.K.; Ono, K.; Kumigashira, H.; et al. Lifting of xz/yz orbital degeneracy at the structural transition in detwinned FeSe. *Phys. Rev. B* **2014**, *90*, 121111. [CrossRef]

27. Zhang, P.; Qian, T.; Richard, P.; Wang, X.P.; Miao, H.; Lv, B.Q.; Fu, B.B.; Wolf, T.; Meingast, C.; Wu, X.X.; et al. Observation of two distinct d_{xz}/d_{yz} band splittings in FeSe. *Phys. Rev. B* **2015**, *91*, 214503. [CrossRef]

28. Song, C.L.; Wang, Y.L.; Cheng, P.; Jiang, Y.P.; Li, W.; Zhang, T.; Li, Z.; He, K.; Wang, L.L.; Jia, J.F.; et al. Direct observation of nodes and twofold symmetry in FeSe Superconductor. *Science* **2011**, *332*, 1410–1413. [CrossRef]

29. Liu, D.; Li, C.; Huang, J.; Lei, B.; Wang, L.; Wu, X.; Shen, B.; Gao, Q.; Zhang, Y.; Liu, X.; et al. Orbital origin of extremely anisotropic superconducting gap in nematic phase of FeSe superconductor. *Phys. Rev. X* **2018**, *8*, 031033. [CrossRef]

30. Mizuguchi, Y.; Tomioka, F.; Tsuda, S.; Yamaguchi, T.; Takano, Y. Superconductivity at 27 K in tetragonal FeSe under high pressure. *Appl. Phys. Lett.* **2008**, *93*, 152505. [CrossRef]

31. Millican, J.N.; Phelan, D.; Thomas, E.L.; Leão, J.B.; Carpenter, E. Pressure-induced effects on the structure of the FeSe superconductor. *Solid State Commun.* **2009**, *149*, 707–710. [CrossRef]

32. Garbarino, G.; Sow, A.; Lejay, P.; Sulpice, A.; Toulemonde, P.; Mezouar, M.; Núñez-Regueiro, M. High-temperature superconductivity (T_c ^{onset} at 34 K) in the high-pressure orthorhombic phase of FeSe. *Europhys. Lett.* **2009**, *86*, 27001. [CrossRef]

33. Imai, T.; Ahilan, K.; Ning, F.L.; McQueen, T.M.; Cava, R.J. Why does undoped FeSe become a high-T_c superconductor under pressure? *Phys. Rev. Lett.* **2009**, *102*, 177005. [CrossRef] [PubMed]

34. Bendele, M.; Ichsanow, A.; Pashkevich, Y.; Keller, L.; Strässle, T.; Gusev, A.; Pomjakushina, E.; Conder, K.; Khasanov, R.; Keller, H. Coexistence of superconductivity and magnetism in $FeSe_{1-x}$ under pressure. *Phys. Rev. B* **2012**, *85*, 064517. [CrossRef]

35. Medvedev, S.; McQueen, T.M.; Troyan, I.A.; Palasyuk, T.; Eremets, M.I.; Cava, R.J.; Naghavi, S.; Casper, F.; Ksenofontov, V.; Wortmann, G.; et al. Electronic and magnetic phase diagram of β-$Fe_{1.01}Se$ with superconductivity at 36.7 K under pressure. *Nat. Mater.* **2009**, *8*, 630–633. [CrossRef]

36. Terashima, T.; Kikugawa, N.; Kasahara, S.; Watashige, T.; Shibauchi, T.; Matsuda, Y.; Wolf, T.; Böhmer, A.E.; Hardy, F.; Meingast, C.; et al. Pressure-Induced antiferromagnetic transition and phase diagram in FeSe. *J. Phys. Soc. Jpn.* **2015**, *84*, 063701. [CrossRef]

37. Bendele, M.; Amato, A.; Conder, K.; Elender, M.; Keller, H.; Klauss, H.H.; Luetkens, H.; Pomjakushina, E.; Raselli, A.; Khasanov, R. Pressure induced static magnetic order in superconducting $FeSe_{1-x}$. *Phys. Rev. Lett.* **2010**, *104*, 087003. [CrossRef]

38. Sun, J.P.; Matsuura, K.; Ye, G.Z.; Mizukami, Y.; Shimozawa, M.; Matsubayashi, K.; Yamashita, M.; Watashige, T.; Kasahara, S.; Matsuda, Y.; et al. Dome-shaped magnetic order competing with high-temperature superconductivity at high pressures in FeSe. *Nat. Commun.* **2016**, *7*, 12146. [CrossRef]

39. Sun, J.P.; Ye, G.Z.; Shahi, P.; Yan, J.Q.; Matsuura, K.; Kontani, H.; Zhang, G.M.; Zhou, Q.; Sales, B.C.; Shibauchi, T.; et al. High-T_c superconductivity in FeSe at high pressure: dominant hole carriers and enhanced spin fluctuations. *Phys. Rev. Lett.* **2017**, *118*, 147004. [CrossRef]

40. Yeh, K.W.; Hsu, C.H.; Huang, T.W.; Wu, P.M.; Huang, Y.L.; Chen, T.K.; Luo, J.L.; Wu, M.K. Se and Te doping study of the FeSe superconductors. *J. Phys. Soc. Jpn.* **2008**, *77*, 19–22. [CrossRef]

41. Yeh, K.W.; Huang, T.W.; Huang, Y.l.; Chen, T.K.; Hsu, F.C.; Wu, P.; Lee, Y.C.; Chu, Y.Y.; Chen, C.L.; Luo, J.Y.; et al. Tellurium substitution effect on superconductivity of the α-phase iron selenide. *Europhys. Lett.* **2008**, *84*, 37002. [CrossRef]

42. Liu, T.J.; Ke, X.; Qian, B.; Hu, J.; Fobes, D.; Vehstedt, E.K.; Pham, H.; Yang, J.H.; Fang, M.H.; Spinu, L.; et al. Charge-carrier localization induced by excess Fe in the superconductor $Fe_{1+y}Te_{1-x}Se_x$. *Phys. Rev. B* **2009**, *80*, 174509. [CrossRef]

43. Yamazaki, T.; Sakurai, T.; Yaguchi, H. Size dependence of oxygen-annealing effects on superconductivity of $Fe_{1+y}Te_{1-x}S_x$. *J. Phys. Soc. Jpn.* **2016**, *85*, 114712. [CrossRef]

44. Sun, Y.; Shi, Z.X.; Tamegai, T. Review of annealing effects and superconductivity in Fe1+yTe1−xSex superconductors. *Supercond. Sci. Technol.* **2019**, *32*, 103001. [CrossRef]

45. Taen, T.; Tsuchiya, Y.; Nakajima, Y.; Tamegai, T. Superconductivity at Tc~14 K in single-crystalline $FeTe_{0.61}Se_{0.39}$. *Phys. Rev. B* **2009**, *80*, 092502. [CrossRef]

46. Sun, Y.; Taen, T.; Tsuchiya, Y.; Shi, Z.X.; Tamegai, T. Effects of annealing, acid and alcoholic beverages on $Fe_{1+y}Te_{0.6}Se_{0.4}$. *Supercond. Sci. Technol.* **2013**, *26*, 015015. [CrossRef]

47. Louca, D.; Horigane, K.; Llobet, A.; Arita, R.; Ji, S.; Katayama, N.; Konbu, S.; Nakamura, K.; Koo, T.Y.; Tong, P.; et al. Local atomic structure of superconducting $FeSe_{1-x}Te_x$. *Phys. Rev. B* **2010**, *81*, 134524. [CrossRef]

48. Liu, T.J.; Hu, J.; Qian, B.; Fobes, D.; Mao, Z.Q.; Bao, W.; Reehuis, M.; Kimber, S.A.; Prokes, K.; Matas, S.; et al. From $(\pi,0)$ magnetic order to superconductivity with (π,π) magnetic resonance in $Fe_{1.02}Te_{1-x}Se_x$. *Nat. Mater.* **2010**, *9*, 716–720. [CrossRef]

49. Mizuguchi, Y.; Tomioka, F.; Tsuda, S.; Yamaguchi, T.; Takano, Y. Superconductivity in S-substituted FeTe. *Appl. Phys. Lett.* **2009**, *94*, 012503. [CrossRef]

50. Abdel, H.M.; Zhang, Y.Y.; Cao, Z.Y.; Duan, C.G.; Karapetrov, G.; Pudalov, V.M.; Vlasenko, V.A.; Sadakov, A.V.; Knyazev, D.A.; Romanova, T.A.; et al. Superconducting properties of sulfur-doped iron selenide. *Phys. Rev. B* **2015**, *91*, 165109. [CrossRef]

51. Watson, M.D.; Kim, T.K.; Haghighirad, A.A.; Blake, S.F.; Davies, N.R.; Hoesch, M.; Wolf, T.; Coldea, A.I. Suppression of orbital ordering by chemical pressure in $FeSe_{1-x}S_x$. *Phys. Rev. B* **2015**, *92*, 121108. [CrossRef]

52. Matsuura, K.; Mizukami, Y.; Arai, Y.; Sugimura, Y.; Maejima, N.; Machida, A.; Watanuki, T.; Fukuda, T.; Yajima, T.; Hiroi, Z.; et al. Maximizing T_c by tuning nematicity and magnetism in $FeSe_{1-x}S_x$ superconductors. *Nat. Commun.* **2017**, *8*, 1143. [CrossRef] [PubMed]

53. Guo, J.G.; Jin, S.; Wang, G.; Wang, S.; Zhu, K.; Zhou, T.; He, M.; Chen, X. Superconductivity in the iron selenide $K_xFe_2Se_2$ ($0 \leq x \leq 1.0$). *Phys. Rev. B* **2010**, *82*, 180520. [CrossRef]

54. Guo, J.G.; Chen, X.L.; Wang, G.; Jin, S.F.; Zhou, T.T.; Lai, X.F. Effect of doping on electrical, magnetic, and superconducting properties of $K_xFe_{2-y}S_2$. *Phys. Rev. B* **2012**, *85*. [CrossRef]

55. Krzton-Maziopa, A.; Shermadini, Z.; Pomjakushina, E.; Pomjakushin, V.; Bendele, M.; Amato, A.; Khasanov, R.; Luetkens, H.; Conder, K. Synthesis and crystal growth of $Cs_{0.8}(FeSe_{0.98})_2$: a new iron-based superconductor with T_c = 27 K. *J. Phys. Condens. Matter.* **2011**, *23*, 052203. [CrossRef] [PubMed]

56. Shermadini, Z.; Krzton-Maziopa, A.; Bendele, M.; Khasanov, R.; Luetkens, H.; Conder, K.; Pomjakushina, E.; Weyeneth, S.; Pomjakushin, V.; Bossen, O.; et al. Coexistence of magnetism and superconductivity in the iron-based compound $Cs_{0.8}(FeSe_{0.98})_2$. *Phys. Rev. Lett.* **2011**, *106*, 117602. [CrossRef] [PubMed]

57. Wang, A.F.; Ying, J.J.; Yan, Y.J.; Liu, R.H.; Luo, X.G.; Li, Z.Y.; Wang, X.F.; Zhang, M.; Ye, G.J.; Cheng, P.; et al. Superconductivity at 32 K in single-crystalline $Rb_xFe_{2-y}Se_2$. *Phys. Rev. B* **2011**, *83*, 060512. [CrossRef]

58. Wang, H.D.; Dong, C.H.; Li, Z.J.; Mao, Q.H.; Zhu, S.S.; Feng, C.M.; Yuan, H.Q.; Fang, M.H. Superconductivity at 32 K and anisotropy in $Tl_{0.58}Rb_{0.42}Fe_{1.72}Se_2$ crystals. *Europhys. Lett. Assoc.* **2011**, *93*, 47004. [CrossRef]

59. Fang, M.H.; Wang, H.D.; Dong, C.H.; Li, Z.J.; Feng, C.M.; Chen, J.; Yuan, H.Q. Fe-based superconductivity with T_c=31 K bordering an antiferromagnetic insulator in $(Tl,K)Fe_xSe_2$. *Europhys. Lett. Assoc.* **2011**, *94*, 27009. [CrossRef]

60. Wang, Z.; Song, Y.J.; Shi, H.L.; Wang, Z.W.; Chen, Z.; Tian, H.F.; Chen, G.F.; Guo, J.G.; Yang, H.X.; Li, J.Q. Microstructure and ordering of iron vacancies in the superconductor system $K_yFe_xSe_2$ as seen via transmission electron microscopy. *Phys. Rev. B* **2011**, *83*, 140505. [CrossRef]

61. Ye, F.; Chi, S.; Bao, W.; Wang, X.F.; Ying, J.J.; Chen, X.H.; Wang, H.D.; Dong, C.H.; Fang, M. Common crystalline and magnetic structure of superconducting $A_2Fe_4Se_5$ (A=K,Rb,Cs,Tl) single crystals measured using neutron diffraction. *Phys. Rev. Lett.* **2011**, *107*, 137003. [CrossRef]

62. Li, W.; Ding, H.; Deng, P.; Chang, K.; Song, C.L.; He, K.; Wang, L.L.; Ma, X.C.; Hu, J.P.; Chen, X.; et al. Phase separation and magnetic order in K-doped iron selenide superconductor. *Nat. Phys.* **2011**, *8*, 126–130. [CrossRef]

63. Li, W.; Ding, H.; Li, Z.; Deng, P.; Chang, K.; He, K.; Ji, S.H.; Wang, L.L.; Ma, X.C.; Hu, J.P.; et al. KFe_2Se_2 is the parent compound of K-doped iron selenide superconductors. *Phys. Rev. Lett.* **2012**, *109*, 057003. [CrossRef] [PubMed]

64. Zhao, J.; Cao, H.; Bourret-Courchesne, E.; Lee, D.H.; Birgeneau, R.J. Neutron-diffraction measurements of an antiferromagnetic semiconducting phase in the vicinity of the high-temperature superconducting state of $K_xFe_{2-y}Se_2$. *Phys. Rev. Lett.* **2012**, *109*, 267003. [CrossRef] [PubMed]

65. Zhao, J.; Shen, Y.; Birgeneau, R.J.; Gao, M.; Lu, Z.Y.; Lee, D.H.; Lu, X.Z.; Xiang, H.J.; Abernathy, D.L.; Zhao, Y. Neutron scattering measurements of spatially anisotropic magnetic exchange interactions in semiconducting $K_{0.85}Fe_{1.54}Se_2$ (T_N = 280 K). *Phys. Rev. Lett.* **2014**, *112*, 177002. [CrossRef] [PubMed]

66. Li, K.K.; Huang, Q.Z.; Zhang, Q.H.; Xiao, Z.W.; Kayami, T.; Hosono, H.; Yuan, D.D.; Guo, J.G.; Chen, X.L. $CsFe_{4-\delta}Se_4$: A compound closely related to alkali-intercalated FeSe superconductors. *Inorg. Chem.* **2018**, *57*, 4502. [CrossRef] [PubMed]

67. Yuan, D.D.; Liu, N.; Li, K.K.; Jin, S.F.; Guo, J.G.; Chen, X.L. Structure evolution and spin-glass transition of layered compounds $ALiFeSe_2$ (A = Na, K, Rb). *Inorg. Chem.* **2017**, *56*, 13187. [CrossRef] [PubMed]

68. Ying, T.P.; Chen, X.L.; Wang, G.; Jin, S.F.; Zhou, T.T.; Lai, X.F.; Zhang, H.; Wang, W.Y. Observation of superconductivity at 30~46 K in $A_x Fe_2 Se_2$ (A = Li, Na, Ba, Sr, Ca, Yb, and Eu). *Sci. Rep.* **2012**, *2*, 426. [CrossRef]
69. Zheng, L.; Izumi, M.; Sakai, Y.; Eguchi, R.; Goto, H.; Takabayashi, Y.; Kambe, T.; Onji, T.; Araki, S.; Kobayashi, T.C.; et al. Superconductivity in $(NH_3)_y Cs_{0.4} FeSe$. *Phys. Rev. B* **2013**, *88*, 094521. [CrossRef]
70. Guo, J.G.; Lei, H.C.; Hayashi, F.; Hosono, H. Superconductivity and phase instability of NH_3-free Na-intercalated $FeSe_{1-z}S_z$. *Nat. Commun.* **2014**, *5*, 4756. [CrossRef]
71. Ying, T.P.; Chen, X.; Wang, G.; Jin, S.; Lai, X.; Zhou, T.; Zhang, H.; Shen, S.; Wang, W. Superconducting phases in potassium-intercalated iron selenides. *J. Am. Chem. Soc.* **2013**, *135*, 2951–2954. [CrossRef]
72. Liu, Y.; Wang, G.; Ying, T.P.; Lai, X.F.; Jin, S.F.; Liu, N.; Hu, J.P.; Chen, X.L. Understanding doping, vacancy, lattice stability, and superconductivity in $K_x Fe_{2-y} Se_2$. *Adv. Sci.* **2016**, 160098. [CrossRef] [PubMed]
73. Burrard-Lucas, M.; Free, D.G.; Sedlmaier, S.J.; Wright, J.D.; Cassidy, S.J.; Hara, Y.; Corkett, A.J.; Lancaster, T.; Baker, P.J.; Blundell, S.J.; et al. Enhancement of the superconducting transition temperature of FeSe by intercalation of a molecular spacer layer. *Nat. Mater.* **2013**, *12*, 15–19. [CrossRef] [PubMed]
74. Jin, S.F.; Fan, X.; Wu, X.Z.; Sun, R.J.; Wu, H.; Huang, Q.Z.; Shi, C.L.; Xi, X.K.; Li, Z.L.; Chen, X.L. High-T_c superconducting phases in organic molecular intercalated iron selenides: synthesis and crystal structures. *Chem. Commun.* **2017**, *53*, 9729–9732. [CrossRef] [PubMed]
75. Fan, X.; Deng, J.; Chen, H.X.; Zhao, L.L.; Sun, R.J.; Jin, S.F.; Chen, X.L. Nematicity and superconductivity in orthorhombic superconductor $Na_{0.35}(C_3 N_2 H_{10})_{0.426} Fe_2 Se_2$. *Phys. Rev. Mater.* **2018**, *2*, 114802. [CrossRef]
76. Lu, X.F.; Wang, N.Z.; Wu, H.; Wu, Y.P.; Zhao, D.; Zeng, X.Z.; Luo, X.G.; Wu, T.; Bao, W.; Zhang, G.H.; et al. Coexistence of superconductivity and antiferromagnetism in $(Li_{0.8} Fe_{0.2})OHFeSe$. *Nat. Mater.* **2015**, *14*, 325–329. [CrossRef]
77. Hayashi, F.; Lei, H.C.; Guo, J.G.; Hosono, H. Modulation effect of interlayer spacing on the superconductivity of electron-doped FeSe-based intercalates. *Inorg. Chem.* **2015**, *54*, 3346–3351. [CrossRef]
78. Hosono, S.; Noji, T.; Hatakeda, T.; Kawamata, T.; Kato, M.; Koike, Y. New intercalation superconductor $Li_x(C_6 H_{16} N_2)_y Fe_{2-z} Se_2$ with a very large interlayer-spacing and $T_c = 38$ K. *J. Phys. Soc. Jpn.* **2014**, *83*, 113704. [CrossRef]
79. Zheng, L.; Sakai, Y.; Miao, X.; Nishiyama, S.; Terao, T.; Eguchi, R.; Goto, H.; Kubozono, Y. Superconductivity in $(NH_3)_y Na_x FeSe_{0.5} Te_{0.5}$. *Phys. Rev. B* **2016**, *94*, 174505. [CrossRef]
80. Sun, L.L.; Chen, X.J.; Guo, J.; Gao, P.W.; Huang, Q.Z.; Wang, H.D.; Fang, M.H.; Chen, X.L.; Chen, G.F.; Wu, Q.; et al. Re-emerging superconductivity at 48 kelvin in iron chalcogenides. *Nature* **2012**, *483*, 67. [CrossRef]
81. Guo, J.; Chen, X.J.; Dai, J.H.; Zhang, C.; Guo, J.G.; Chen, X.L.; Wu, Q.; Gu, G.D.; Gao, P.W.; Yang, L.H.; et al. Pressure-driven quantum criticality in iron-selenide superconductors. *Phys. Rev. Lett.* **2012**, *108*, 197001. [CrossRef]
82. Sun, J.P.; Shahi, P.; Zhou, H.X.; Huang, Y.L.; Chen, K.Y.; Wang, B.S.; Ni, S.L.; Li, N.N.; Zhang, K.; Yang, W.G.; et al. Reemergence of high-T_c superconductivity in the $(Li_{1-x} Fe_x)OHFe_{1-y} Se$ under high pressure. *Nat. Commun.* **2018**, *9*, 380. [CrossRef]
83. Shahi, P.; Sun, J.P.; Sun, S.S.; Jiao, Y.Y.; Chen, K.Y.; Wang, S.H.; Lei, H.C.; Uwatoko, Y.; Wang, B.S.; Cheng, J.-G. High-T_c superconductivity up to 55 K under high pressure in the heavily electron doped $Li_x(NH_3)_y Fe_2 Se_2$ single crystal. *Phys. Rev. B* **2018**, *97*, 020508. [CrossRef]
84. Qian, T.; Wang, X.P.; Jin, W.C.; Zhang, P.; Richard, P.; Xu, G.; Dai, X.; Fang, Z.; Guo, J.G.; Chen, X.L.; et al. Absence of a holelike fermi surface for the iron-based $K_{0.8} F_{1.7} Se_2$ superconductor revealed by angle-resolved photoemission spectroscopy. *Phys. Rev. Lett.* **2011**, *106*, 187001. [CrossRef] [PubMed]
85. Zhao, L.; Liang, A.J.; Yuan, D.N.; Hu, Y.; Liu, D.; Huang, J.W.; He, S.L.; Shen, B.; Xu, Y.; Liu, X.; et al. Common electronic origin of superconductivity in (Li,Fe)OHFeSe bulk superconductor and single-layer $FeSe/SrTiO_3$ films. *Nat. Commun.* **2016**, *7*, 10608. [CrossRef] [PubMed]
86. Zhao, L.; Mou, D.X.; Liu, S.Y.; Jia, X.W.; He, J.F.; Peng, Y.Y.; Yu, L.; Liu, X.; Liu, G.D.; He, S.L.; et al. Common Fermi-surface topology and nodeless superconducting gap of $K_{0.68} Fe_{1.79} Se_2$ and $(Tl_{0.45} K_{0.34})Fe_{1.84} Se_2$ superconductors revealed via angle-resolved photoemission. *Phys. Rev. B* **2011**, *83*, 140508. [CrossRef]
87. Mazin, I.I.; Singh, D.J.; Johannes, M.D.; Du, M.H. Unconventional superconductivity with a sign reversal in the order parameter of $LaFeAsO_{1-x} F_x$. *Phys. Rev. Lett.* **2008**, *101*, 057003. [CrossRef] [PubMed]

88. Zhang, Y.; Yang, L.X.; Xu, M.; Ye, Z.R.; Chen, F.; He, C.; Xu, H.C.; Jiang, J.; Xie, B.P.; Ying, J.J.; et al. Nodeless superconducting gap in $A_xFe_2Se_2$ (A=K,Cs) revealed by angle-resolved photoemission spectroscopy. *Nat. Mater.* **2011**, *10*, 273–277. [CrossRef] [PubMed]

89. Wang, Q.Y.; Li, Z.; Zhang, W.H.; Zhang, Z.C.; Zhang, J.S.; Li, W.; Ding, H.; Ou, Y.B.; Deng, P.; Chang, K.; et al. Interface-induced high-temperature superconductivity in single unit-cell FeSe films on $SrTiO_3$. *Chin. Phys. Lett.* **2012**, *29*, 037402. [CrossRef]

90. Ge, J.F.; Liu, Z.L.; Liu, C.H.; Gao, C.L.; Qian, D.; Xue, Q.K.; Liu, Y.; Jia, J.F. Superconductivity above 100 K in single-layer FeSe films on doped $SrTiO_3$. *Nat. Mater.* **2015**, *14*, 285–289. [CrossRef]

91. He, S.L.; He, J.F.; Zhang, W.H.; Zhao, L.; Liu, D.F.; Liu, X.; Mou, D.X.; Ou, Y.B.; Wang, Q.Y.; Li, Z. Phase diagram and electronic indication of high-temperature superconductivity at 65 K in single-layer FeSe films. *Nat. Mater.* **2013**, *12*, 605–610. [CrossRef]

92. Tan, S.Y.; Zhang, Y.; Xia, M.; Ye, Z.R.; Chen, F.; Xie, X.; Peng, R.; Xu, D.F.; Fan, Q.; Xu, H.C. Interface-induced superconductivity and strain-dependent spin density waves in $FeSe/SrTiO_3$ thin films. *Nat. Mater.* **2013**, *12*, 634–640. [CrossRef] [PubMed]

93. Rebec, S.N.; Jia, T.; Zhang, C.; Hashimoto, M.; Lu, D.H.; Moore, R.G.; Shen, Z.X. Coexistence of replica bands and superconductivity in FeSe monolayer films. *Phys. Rev. Lett.* **2017**, *118*, 067002. [CrossRef] [PubMed]

94. Lee, J.J.; Schmitt, F.T.; Moore, R.G.; Johnston, S.; Cui, Y.T.; Li, W.; Yi, M.; Liu, Z.K.; Hashimoto, M.; Zhang, Y. Interfacial mode coupling as the origin of the enhancement of T_c in FeSe films on $SrTiO_3$. *Nature* **2014**, *515*, 245–248. [CrossRef] [PubMed]

95. Ding, H.; Lv, Y.F.; Zhao, K.; Wang, W.L.; Wang, L.L.; Song, C.L.; Chen, X.; Ma, X.C.; Xue, Q.K. High-temperature superconductivity in single-unit-cell FeSe films on anatase $TiO_2(001)$. *Phys. Rev. Lett.* **2016**, *117*. [CrossRef]

96. Zhou, G.Y.; Zhang, Q.H.; Zheng, F.W.; Zhang, D.; Liu, C.; Wang, X.X.; Song, C.L.; He, K.; Ma, X.C.; Gu, L. Interface enhanced superconductivity in monolayer FeSe films on MgO (001): charge transfer with atomic substitution. *Sci. Bull.* **2019**, *63*, 747–752. [CrossRef]

97. Miyata, Y.; Nakayama, K.; Sugawara, K.; Sato, T.; Takahashi, T. High-temperature superconductivity in potassium-coated multilayer FeSe thin films. *Nat. Mater.* **2015**, *14*, 775–779. [CrossRef]

98. Song, Q.; Yu, T.L.; Lou, X.; Xie, B.P.; Xu, H.C.; Wen, C.H.P.; Yao, Q.; Zhang, S.Y.; Zhu, X.T.; Guo, J.D.; et al. Evidence of cooperative effect on the enhanced superconducting transition temperature at the $FeSe/SrTiO_3$ interface. *Nat. Commun.* **2019**, *10*, 758. [CrossRef]

99. Rodriguez, J.P. Isotropic Cooper pairs with emergent sign changes in a single-layer iron superconductor. *Phys. Rev. B* **2017**, *95*, 134511. [CrossRef]

100. Rodriguez, J.P.; Melendrez, R. Fermi surface pockets in electron-doped iron superconductor by Lifshitz transition. *J. Phys. Commun.* **2018**, *2*, 105011. [CrossRef]

 © 2019 by the authors. Licensee MDPI, Basel, Switzerland. This article is an open access article distributed under the terms and conditions of the Creative Commons Attribution (CC BY) license (http://creativecommons.org/licenses/by/4.0/).

Article

Comparative Study on AC Susceptibility of YBa$_2$Cu$_3$O$_{7-\delta}$ Added with BaZrO$_3$ Nanoparticles Prepared via Solid-State and Co-Precipitation Method

Nurhidayah Mohd Hapipi [1], Jee Khan Lim [2], Soo Kien Chen [1,3,*], Oon Jew Lee [2], Abdul Halim Shaari [1], Mohd Mustafa Awang Kechik [1], Kean Pah Lim [1], Kar Ban Tan [4], Masato Murakami [5] and Muralidhar Miryala [5]

[1] Department of Physics, Faculty of Science, Universiti Putra Malaysia, 43400 UPM Serdang, Selangor, Malaysia; gs52310@student.upm.edu.my (N.M.H.); ahalim@upm.edu.my (A.H.S.); mmak@upm.edu.my (M.M.A.K.); limkp@upm.edu.my (K.P.L.)
[2] School of Fundamental Science, Universiti Malaysia Terengganu, 21030 Kuala Nerus, Terengganu, Malaysia; p3595@pps.umt.edu.my (J.K.L.); oonjew@umt.edu.my (O.J.L.)
[3] Institute of Advanced Technology, Universiti Putra Malaysia, 43400 UPM Serdang, Selangor, Malaysia
[4] Department of Chemistry, Faculty of Science, Universiti Putra Malaysia, 43400 UPM Serdang, Selangor, Malaysia; tankarban@upm.edu.my
[5] Shibaura Institute of Technology, 3 Chome-7-5 Toyosu, Koto, Tokyo 135-8548, Japan; masatomu@shibaura-it.ac.jp (M.M.); miryala1@shibaura-it.ac.jp (M.M.)
* Correspondence: chensk@upm.edu.my

Received: 11 September 2019; Accepted: 6 November 2019; Published: 9 December 2019

Abstract: Polycrystalline samples of YBa$_2$Cu$_3$O$_{7-\delta}$ (Y-123) added with x mol% of BaZrO$_3$ (BZO) nanoparticles (x = 0.0, 2.0, 5.0, and 7.0) were synthesized using co-precipitation (COP) and solid-state (SS) method. X-ray diffraction (XRD) patterns showed the formation of Y-123 and Y-211 as the major and minor phases, respectively. The samples prepared using COP method showed higher weight percentage of Y-123 phase ($\leq$98%) compared to the SS samples ($\leq$93%). A peak corresponding to BZO was also found in the samples added with BZO nanoparticles. The increasing intensity of the BZO peak as the BZO amount increased showed the increasing amount of the unreacted nanoparticles in the samples. Refinement of unit cell lattice parameters indicated that all the samples have an orthorhombic crystal structure and there is no orthorhombic-tetragonal phase transformation. As observed using scanning electron microscopy (SEM), all the samples showed randomly distributed grains with irregular shape. The average grain size for the pure sample prepared using COP method is smaller (0.30 μm) compared with that of the pure SS sample (1.24 μm). Addition of 7.0 mol% BZO led to an increase of average grain size to 0.50 μm and 2.71 μm for the COP and SS samples, respectively, indicating grain growth. AC susceptibility (ACS) measurement showed a decrease in the onset critical temperature, $T_{\text{c-onset}}$ with BZO addition. Comparatively, $T_{\text{c-onset}}$ for the COP samples is higher than that of the SS samples. The value of Josephson's current, I_o increased up to 2.0 mol% BZO addition, above which the I_o decreased more drastically for the SS samples. The value of I_o is 53.95 μA and 32.08 μA for the 2.0 mol% BZO added SS and COP samples, respectively. The decrease of I_o is attributed to the distribution of BZO particles at the grain boundaries as also reflected in the drastic decrease of phase lock-in temperature, T_{cj}. As a result of smaller average grain size, the presence of more grain boundaries containing insulating BZO particles led to lower I_o in the COP samples.

Keywords: AC susceptibility; BaZrO$_3$; co-precipitation; solid-state; YBa$_2$Cu$_3$O$_{7-\delta}$

1. Introduction

Yttrium-barium-copper-oxide, YBCO is the first type-II high temperature superconductor (HTS) that has been discovered to be superconducting above the boiling point of liquid nitrogen. Among the family members of YBCO, 123 phase (Y-123) shows the highest onset of superconductivity at around 90 K. In order for Y-123 to be more feasible for a wide range of applications, numerous studies have been undertaken to improve its critical temperature, T_c and critical current density, J_c [1–3]. In this regard, addition of chemical dopants is one of the most straight forward strategies. It was shown that the addition of $BaZrO_3$ (BZO) nanoparticles into Y-123 improved J_c without affecting much the T_c value [4,5]. This is because BZO did not dope into the structure of Y-123. Instead, they either remained unreacted within the matrix or reacted with Y-123 to form nano-precipitates leading to enhanced flux pinning [6].

Intensive research on fabrication of BZO doped YBCO films was conducted in the past [7–11]. However, limited study of BZO doped YBCO bulks has been reported so far [5,12]. From the perspective of large scale applications, it is essential to investigate the effects of synthesis route and dopant additions on the superconducting properties of YBCO bulks. There are several methods used to synthesize the materials. For example, solid-state (SS) method [13,14] and wet method such as sol-gel [15], co-precipitation (COP) [16–18], and thermal treatment [19]. In particular, COP is a highly desirable chemical method used to prepare nanomaterials. The powders obtained by COP method show smaller grain size, higher purity, and better homogeneity compared to that obtained using SS method [20,21]. Moreover, multiple calcination and regrinding process is unnecessary to obtain a good superconducting phase for the COP method [17,22]. Previously, we successfully synthesized YBCO added with nano BZO using COP method [23]. Therefore, we are motivated to carry out a comparative study on structural and superconducting properties of BZO nanoparticles added Y-123 prepared via SS method and COP method. COP can be defined as the process of carrying down a precipitate of substances which are normally soluble under the suitable conditions [24] and the starting materials can be metal acetates or metal nitrates [14].

2. Materials and Methods

Samples with nominal composition of $YBa_2Cu_3O_{7-\delta}$ (Y-123) added with x mol% of $BaZrO_3$ (BZO) nanoparticles (x = 2.0, 5.0, and 7.0) were prepared using solid state (SS) method and co-precipitation (COP) method, respectively.

2.1. Solid-State (SS) Method

To start with, Y_2O_3 (99.9%, Alfa Aesar, Ward Hill, Massachusetts, USA), $BaCO_3$ (99.8%, Alfa Aesar, Ward Hill, Massachusetts, USA) and CuO (99.9%, Strem Chemical, Newburyport, Massachusetts, USA) with the stoichiometric ratio of Y:Ba:Cu (1:2:3) were mixed and hand-ground for 1 h using a mortar and pestle. Then, the grounded mixture was calcined in air at 940 °C for 12 h. After calcination, the powders were reground for 1 h before adding x mol% (x = 2.0, 5.0, and 7.0, respectively) of $BaZrO_3$ nanoparticles (BZO, >50 nm, 98.5%, Sigma-Aldrich). The mixed powders were reground again and then pressed into circular pellets (~13-mm diameter and 2-mm thickness) by using a hydraulic press with an applied pressure load of five tons. Finally, the pellets were sintered at 950 °C for 12 h and slowly cooled to 450 °C for 12 h at the rate of 1 °C/min under oxygen flow before further cooling to room temperature. Pure samples (x = 0.0) were also prepared according to the same procedure to serve as reference for the purpose of comparison.

2.2. Co-Precipitation (COP) Method

For COP method, appropriate amounts of $Y(CH_3COO)_3 \cdot 4H_2O$ (99.9% Alfa Aesar), $Ba(CH_3COO)_2$ ($\geq$99% Alfa Aesar), and $Cu(CH_3COO)_2 \cdot H_2O$ ($\geq$99% Sigma Aldrich) according to the stoichiometric ratio of Y:Ba:Cu (1:2:3) were dissolved in acetic acid to form solution A. To prepare solution B, oxalic acid was dissolved in a mixture of distilled water: 2-propanol (1:1.5). Both solutions A and B were stirred at 300 rpm for 2 h before being cooled in an ice bath. The mixed solution A and B was filtered and dried at 100 °C for 12 h. The obtained dried powders were ground and calcined in air at 900 °C for 24 h. After that, appropriate amount of x mol% (x = 2.0, 5.0, and 7.0, respectively) $BaZrO_3$ nanoparticles (BZO, >50 nm, 98.5%, Sigma-Aldrich) was added to the calcined powders for mixing and grinding. Then, the mixture was pressed into circular pellets (~13-mm diameter and 2-mm thickness) by using a hydraulic press with an applied pressure load of 5 tons. Lastly, the pellets were sintered at 920 °C for 15 h and slowly cooled to 650 °C for 8 h (annealing process) before further cooling to room temperature. The sintering and annealing were done under a constant oxygen flow. Pure samples (x = 0.0) were also prepared according to the same procedure to serve as reference for the purpose of comparison.

2.3. Sample Characterization

Phase formation and crystal structure of the samples were examined by X-ray diffraction (XRD) method using the PW 3040/60 MPD X'Pert Pro Panalytical Philips DY 1861 X-ray diffractometer with Cu-K$_\alpha$ radiation source. Scanning was carried out in 2θ mode over the range of 20°–80° with the increment step size of 0.03°. The XRD data was analyzed using the X'pert HighScore Plus software. Surface morphology of the pellets was observed using a scanning electron microscope (SEM-LEO 1455 VPSEM). Superconducting properties of the samples were measured using the commercial AC Susceptometer of CryoBIND (cryogenic balanced inductive detector) SR830 at the frequency of 219 Hz and applied field of 0.5 Oe. Uncertainty of the AC Susceptometer is ±0.1 K.

3. Discussion

3.1. X-ray Diffraction (XRD) Analysis

Figure 1 shows the XRD patterns of the samples prepared via SS method and COP method, respectively. The XRD patterns were indexed to Y-123 phase with orthorhombic crystal structure and space group *Pmmm* (ICSD: 01-078-2143). Peaks with the highest intensity for both the SS and COP samples were indexed with the Miller indices of (0 1 3) and (1 0 3). Rietveld refinement was undertaken on the XRD data using the X'pert HighScore Plus software. Accordingly, the refined unit cell lattice parameters indicated that all the samples have an orthorhombic crystal structure. Hence, no structural transformation from orthorhombic to tetragonal phase was occurred as a result of BZO addition. By using the similar refinement method, fraction of different phases (in weight percentage) in the samples was obtained. Compared to SS method, the samples prepared using COP method have higher weight percentage of Y-123 phase in agreement with previous findings [20,21]. Table 1 indicates the weight percentage of Y-123 phase for the pure SS and COP samples is 93.2% and 97.5%, respectively. However, the weight percentage of Y-123 phase for the SS samples and COP samples started to decrease with the increasing of BZO additions. Some minor peaks of Y_2BaCuO_5 (Y-211) (ICSD: 01-079-0697) were also observed in the XRD patterns of all the samples. The presence of Y-211 phase may due to the synthesis condition such as the heat treatment process and the preparation of the material itself [25]. A peak corresponding to $BaZrO_3$ (BZO) (ICSD: 01-074-1299) started to appear at 2.0 mol% BZO addition and became more intense with increasing of BZO addition. This observation confirmed the distribution of BZO in the samples [26]. The increased intensity of the peak with increasing BZO addition suggests the saturation of solid solubility of BZO in Y-123 [27].

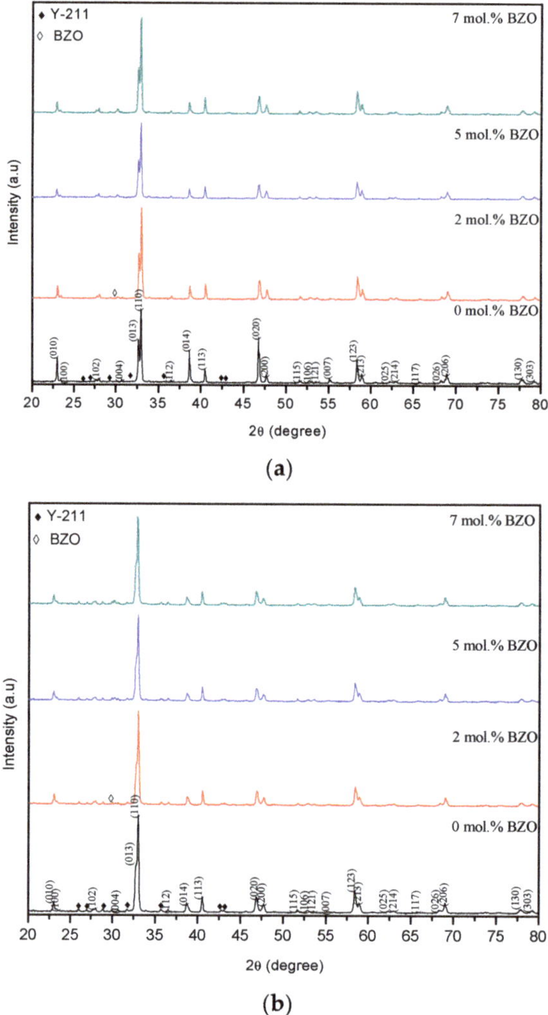

Figure 1. X-ray diffraction patterns of Y-123 + *x* mol% of BZO samples prepared via (**a**) SS method and (**b**) COP method. The Y-211 and BZO are marked with (♦) and (◊), respectively.

Table 1. Percentages of Y-123, Y-211, and BZO phases for Y-123 + *x* mol% of BZO samples prepared via SS method and COP method.

BZO *x* (mol%)	Weight Percentage of Phases (%)					
	SS Method			COP Method		
	Y-123	Y-211	BZO	Y-123	Y-211	BZO
0.0	93.2	6.8	0.0	97.5	2.5	0.0
2.0	92.9	6.6	0.5	96.6	3.1	0.3
5.0	92.8	5.7	1.5	94.8	4.1	2.7
7.0	92.8	3.8	3.4	94.3	3.0	2.7

Table 2 summarizes crystallite size calculated based on the (103) peak of Y-123 using the Scherrer equation [28]:

$$L = \frac{K\lambda}{B_{size}\cos\theta}$$

(1)

where L is the crystallite size, K is a dimensionless shape factor (0.9), B_{size} is line broadening at half of the maximum intensity (FWHM) in radian, λ is the X-ray wavelength for Cu-K$_\alpha$ radiation (1.5406 Å) and θ is Bragg angle in degree. Table 2 shows that the crystallite size of the SS samples is larger than that of the COP samples. Crystallite size of the pure SS and COP samples is 101.61 nm and 63.66 nm, respectively. Upon addition of BZO, the crystallite size increased for the COP samples. The larger crystallite size may be due to the bridging of fine particles that formed the continuous grain boundary networks [29]. Conversely, crystallite size became smaller for the SS samples with addition of BZO. Previously, the crystalline size of $CaTiO_3$ (CTO) was found to decrease possibly due to large difference in ionic radii between the dopant ion and host ion (Ti^{4+}) [30]. The nanocrystalline CTO was synthesized using soft chemical method followed by annealing. For our work, however, the same BZO nanoparticles were used for addition into the samples prepared by SS method and COP method. Judging from the data given in Table 2, the decrease in crystalline size for the BZO added SS samples is believed to be due to larger difference of the crystalline size between the BZO and Y-123 compared to that between the BZO and Y-123 prepared using COP method.

Table 2. FWHM of (103) peak, crystallite size and average grain size for Y-123 + *x* mol% of BZO samples calculated using Scherrer method. The average grain size of the samples was calculated from the randomly selected 100 grains of the SEM image.

	SS Method			COP Method		
BZO *x* (mol%)	FWHM of (103) Peak (°)	Crystallite Size (nm)	Average Grain Size, D (µm)	FWHM of (103) Peak (°)	Crystallite Size (nm)	Average Grain Size, D (µm)
0.0	0.1315	101.61	1.24 ± 0.14	0.1801	63.66	0.30 ± 0.01
2.0	0.1424	89.62	1.77 ± 0.06	0.1737	66.94	0.39 ± 0.02
5.0	0.1515	81.56	2.29 ± 0.01	0.1793	64.03	0.45 ± 0.02
7.0	0.1436	88.44	2.71 ± 0.11	0.1729	67.36	0.02

3.2. Microstructure Analysis

Figure 2 shows the SEM images of the SS and COP samples. All the samples have irregular shape and randomly distributed grains. The SEM images also show the increment in number of pores with BZO addition suggesting more porous structure for both SS and COP samples added with BZO. For calculation of average grain size, 100 grains were selected randomly from each SEM image and measured using the Image-J software. Figure 3 shows the distribution of grain size of the SS and COP samples. The average grain size is 1.24 µm and 0.30 µm, for the pure samples prepared via the SS method and COP method, respectively (Table 2). Addition of BZO into Y-123 sample increased the average grain size of the samples especially for the SS samples. The average grain size increased to 2.71 µm and 0.50 µm for the 7.0 mol% BZO added SS and COP samples, respectively. The increasing grain size is probably due to dispersion of BZO nanoparticles between the grains of Y-123 which promoted grain growth [5,31]. According to Table 2, the powders obtained by the COP method have smaller size and narrower size range (0.30–0.50 µm) in comparison with that obtained via the SS method (1.24–2.71 µm). This may be partly associated with the higher sintering temperature used in the SS method (950 °C, compared with 920 °C used for the COP method) leading to larger grain growth [21].

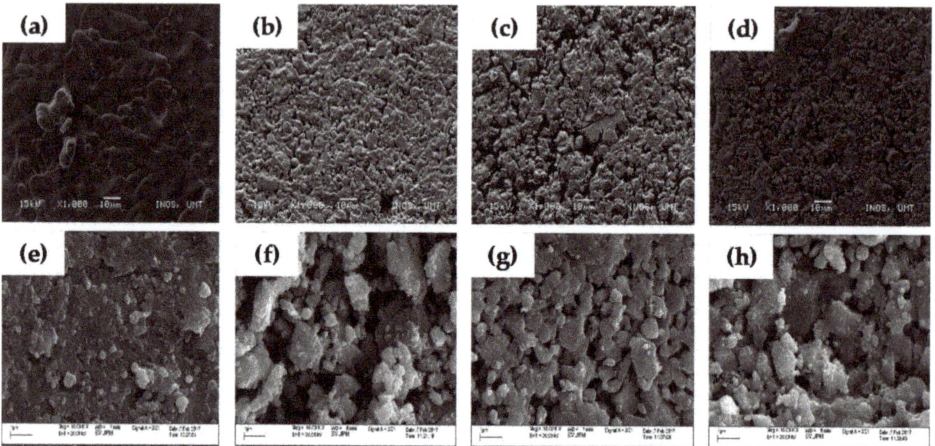

Figure 2. SEM images of Y-123 added with (**a**) 0.0 mol%, (**b**) 2.0 mol%, (**c**) 5.0 mol%, and (**d**) 7.0 mol% of BZO at 5000× magnification prepared via SS method. Bottom: SEM images of Y-123 added with (**e**) 0.0 mol%, (**f**) 2.0 mol%, (**g**) 5.0 mol%, and (**h**) 7.0 mol% of BZO at 5000× magnification prepared via COP method.

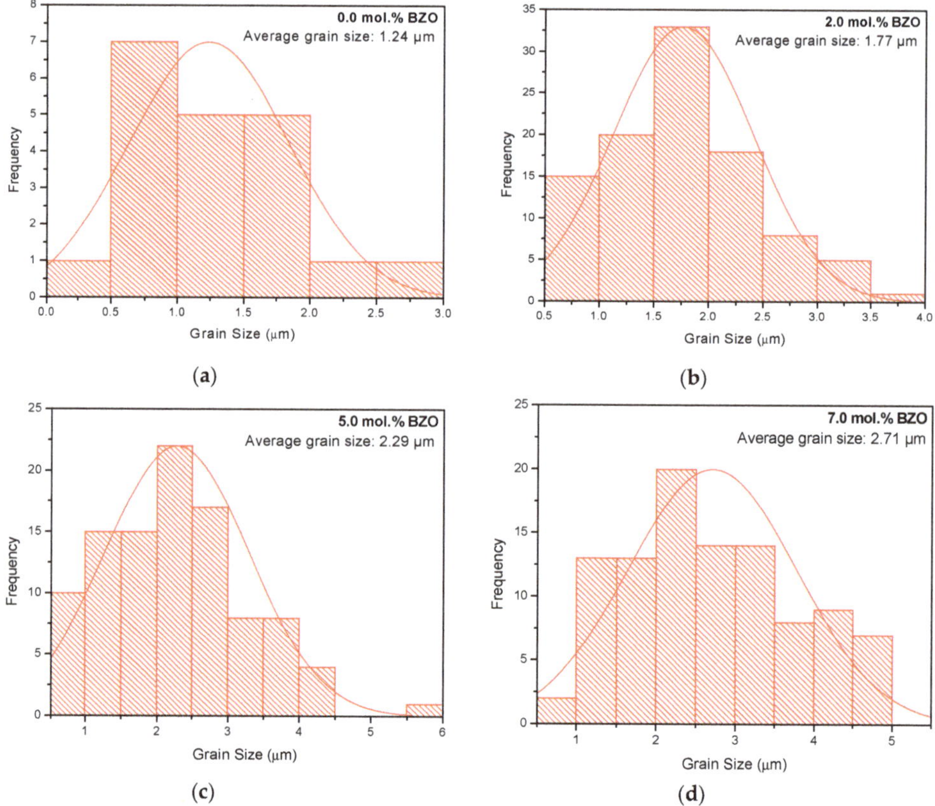

(**a**)   (**b**)

(**c**)   (**d**)

Figure 3. *Cont.*

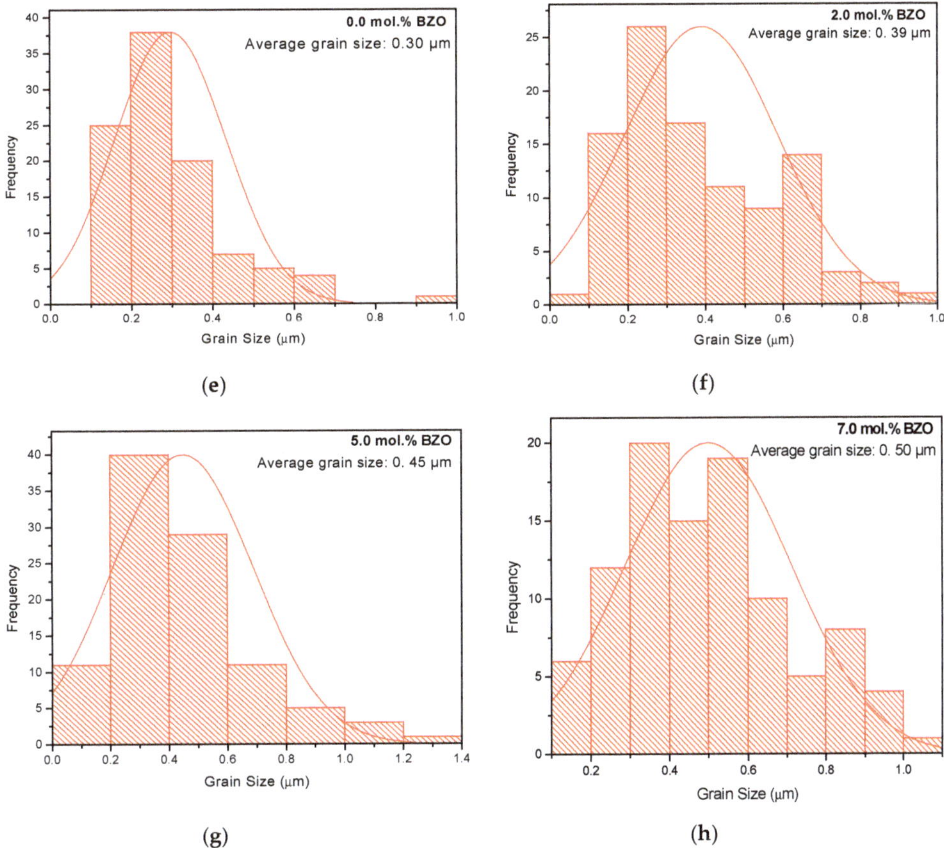

Figure 3. Distribution of average grain size for Y-123 added with (**a**) 0.0 mol%, (**b**) 2.0 mol%, (**c**) 5.0 mol%, and (**d**) 7.0 mol% of BZO prepared via SS method and Y-123 added with (**e**) 0.0 mol%, (**f**) 2.0 mol%, (**g**) 5.0 mol%, and (**h**) 7.0 mol% of BZO prepared via COP method.

3.3. Superconducting Properties

Figure 4 shows the temperature dependence of the real (χ') and imaginary (χ'') parts of AC susceptibility (ACS) for all the samples. The real part (χ') shows two transitions labeled as onset critical temperature, $T_{c\text{-onset}}$ and phase lock-in temperature, T_{cj}. $T_{c\text{-onset}}$ is due to transition within the grains (intra-grain) while T_{cj} is due to superconducting coupling between grains (inter-grain) [32]. In the temperature range between $T_{c\text{-onset}}$ and T_{cj}, the superconducting grains are decoupled and the sample as a whole becomes resistive even if all individual grains are still strongly superconducting. Below T_{cj}, the grains are coupled or phase-locked in which the phase difference across the intergranular junctions is zero. Consequently, the induced current from the externally applied ac magnetic field gives rise to shielding currents along the outermost surface of the sample. $T_{c\text{-onset}}$ and T_{cj} were determined from the peaks of the derivative of normalized χ', $\delta\chi'/\delta T$ against temperature, T plots (insets). It should also be noted that samples with approximately the same size (cross-section 0.3×0.5 cm) were used for measurements under the same conditions for the sake of consistency.

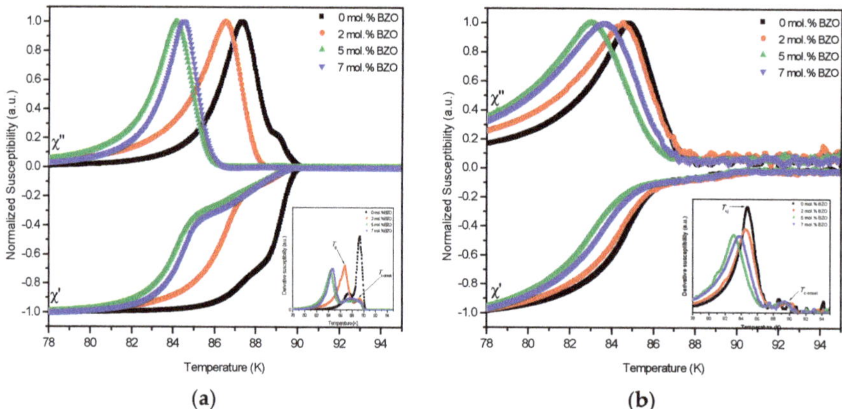

Figure 4. Temperature dependence of ACS for Y-123 + x mol% of BZO samples prepared via (**a**) SS method (**b**) COP method. Insets show the derivative of normalized χ', $\delta\chi'/\delta T$ against temperature, T.

Table 3 shows that the values of $T_{\text{c-onset}}$ and T_{cj} for the COP samples are close to that of the SS samples. The difference for $T_{\text{c-onset}}$ and T_{cj} is ≤1.6 K and ≤1.2 K, respectively. With increasing of BZO addition from 0.0 mol% to 7.0 mol%, $T_{\text{c-onset}}$ and T_{cj} for the COP samples decreased slightly (0.7 K and 0.8 K for $T_{\text{c-onset}}$ and T_{cj}, respectively). Nevertheless, addition of 2 mol% BZO did not change the value of T_{cj}. The marginal decrease of $T_{\text{c-onset}}$ is in good agreement with the previous finding [12,33,34] and could be associated with the reduced hole concentration [23]. However, the values of $T_{\text{c-onset}}$ and T_{cj} for the SS samples showed a larger change compared to the COP samples. For the SS samples, $T_{\text{c-onset}}$ decreased from 90.3 K for 0.0 mol% to 88.9 K for 7.0 mol% BZO addition (1.4 K). The lower $T_{\text{c-onset}}$ of the 7.0 mol% BZO added SS sample compared to that of the COP sample may be attributed to inhomogeneity of the former. Meanwhile, the value of T_{cj} decreased from 87.8 K for 0.0 mol% to 85.5 K for 7.0 mol% BZO addition (2.3 K). The decrease of T_{cj} with the increasing of BZO addition for both COP and SS samples indicates weakening of intergranular coupling perhaps due to the increasing of impurities at the grain boundaries [35,36]. However, the larger decrease of T_{cj} in the SS samples implies more severe degradation of grain coupling caused by BZO addition compared to the COP samples.

Table 3. Coupling peak temperature, T_p, onset critical temperature, $T_{\text{c-onset}}$, phase lock-in temperature, T_{cj}, and Josephson current, I_o for Y-123 + x mol% of BZO samples in applied AC field of 0.5 Oe.

BZO x (mol%)	SS Method				COP Method			
	T_p (K)	$T_{\text{c-onset}}$ (K)	T_{cj} (K)	I_o (μA)	T_p (K)	$T_{\text{c-onset}}$ (K)	T_{cj} (K)	I_o (μA)
0.0	87.3	90.3	87.8	51.21	84.8	90.6	86.6	31.69
2.0	85.8	88.9	86.6	53.95	84.5	90.5	86.5	32.08
5.0	84.1	89.9	85.4	28.20	82.9	89.8	85.6	30.56
7.0	84.5	88.9	85.5	36.49	83.6	89.9	85.8	30.54

Imaginary part (χ'') of ACS exhibits a broad peak known as coupling peak temperature, T_p that corresponds to the hysteretic energy dissipation as a result of AC losses at the inter-grain region [37]. The second peak near T_c (intra-granular peak) for all the samples is not apparent except for the pure sample prepared via the SS method. The absence of the second peak suggests that grain coupling is better for the pure sample prepared via COP method compared to the SS method [37,38]. For the COP samples, addition of 2.0 mol% of BZO did not change the T_p much compared to the pure sample. For the SS samples, T_p for the pure sample is 87.3 K compared to 85.8 K for the sample with 2.0 mol% of BZO addition. As shown in Figure 4 and Table 3, T_p of both the SS and COP samples shifted to lower temperature with the increasing of BZO addition due to the weakening of intergranular coupling between the grains [39]. Other factors that affect T_p also include the changes in microstructure and the amount of impurities at the grain boundaries [40]. It is noteworthy that both T_p and T_{cj} of the 5.0 and 7.0 mol% BZO added samples are quite close to each other for the COP and SS samples. Such a small variation could be related to the saturation of solid solubility of BZO in Y-123 as mentioned in the part of XRD analysis (Section 3.1).

The value of Josephson's current, I_o was calculated according to the Ambegaokar–Baratoff theory [41] using the following equation assuming that nearest-neighbor grains are coupled by identical Josephson junctions of maximum I_o:

$$I_o = 1.57 \times 10^{-8} \left(\frac{T_{\text{c-onset}}^2}{T_{\text{c-onset}} - T_{cj}} \right) \tag{2}$$

It should be noted that Equation (2) is only valid for s-wave superconductors at temperature near to the superconducting transition temperature, T_c. Nonetheless, the equation has been adapted for the estimation of maximum Josephson current in d-wave cuprate superconductor [42]. The values of $T_{\text{c-onset}}$ and T_{cj} were estimated from Figure 4 as mentioned before. Due to the granular nature of HTS, grain boundaries of Y-123 act as Josephson junctions. In equation (2), T_{cj} represents the Josephson phase-locking temperature which is the crossover temperature between phase-locked and phase fluctuation dominated behavior, corresponding to coupling and decoupling of grains, respectively [41].

Table 3 shows that the I_o increased for 2.0 mol% samples but decreased for 5.0 mol% and 7.0 mol% BZO addition for both methods. The Josephson current, I_o = 53.95 µA for 2 mol% BZO added SS sample is the highest. Meanwhile, I_o for 2.0 mol% BZO added COP sample is 32.08 µA. These values are higher than that $I_o \approx 5$ µA in Bi-Ca-Sr-Cu-O bulk prepared using solid state reaction method [42]. The increasing of Josephson's current, I_o up to 2.0 mol% indicated better coupling between the grains and thus, stronger Josephson junction. Consequently, the screening current circulated even at higher field causing the increasing tunneling of Josephson current across the grain boundaries [43]. However, further addition of 5.0 mol% and 7.0 mol% BZO in the SS samples, abruptly decreased the I_o to 28.20 µA and 36.49 µA, respectively. For the COP samples, I_o decreased to 30.56 µA and 30.54 µA with addition of 5.0 and 7.0 mol% BZO, respectively. Nevertheless, the variation in I_o with increased BZO addition in the COP samples is smaller probably because of higher homogeneity over the SS samples. Such decrease in I_o is reflected in the more drastic decrease of T_{cj} of the 5.0 and 7.0 mol% BZO added samples (Table 3). This is most probably related to the distribution of BZO particles which are insulating at the grain boundaries. The same reason is believed to be accounted for the lower I_o of the COP samples compared to that of the SS samples. This can be understood by the presence of more grain boundaries in the COP samples due to smaller grain size (Table 2). Hence, tunneling of Josephine current is hindered due to the distribution of the insulating BZO particles at the grain boundaries.

4. Conclusions

In this work, structural properties of BZO doped Y-123 bulks prepared using COP and SS method, respectively, were studied and their AC susceptibility was compared. The polycrystalline samples with nominal composition of Y-123 added with x mol% of BZO nanoparticles (x = 0.0, 2.0, 5.0, and 7.0) were synthesized. The XRD patterns showed the formation of Y-123 as the major phase and Y-211 as the secondary phase. The COP samples showed higher weight percentage of Y-123 phase. Refinement of unit cell lattice parameters indicated that all the samples have an orthorhombic crystal structure. Estimation based on the SEM images showed that the COP samples have smaller average grain size than that of the SS samples probably because of lower sintering temperature used for the former. $T_{\text{c-onset}}$ was found to decrease with addition of BZO as shown by the ACS measurement. The value of Josephson's current, I_o increased up 53.95 µA and 32.80 µA for 2.0 mol% BZO added SS and COP samples, respectively. The lower I_o value of the COP samples compared with the SS samples is attributed to the presence of more grain boundaries. Hence, tunneling of I_o is hindered due to the distribution of insulating BZO particles at the grain boundaries.

Author Contributions: Conceptualization: N.M.H., J.K.L., S.K.C., and O.J.L.; methodology: N.M.H., J.K.L., S.K.C., and O.J.L; validation: N.M.H., J.K.L., S.K.C., O.J.L., A.H.S., M.M.A.K., K.P.L., K.B.T., M.M., and M.M.; writing—original draft preparation: N.M.H.; writing—review and editing: J.K.L., S.K.C., and O.J.L; supervision: S.K.C., O.J.L., M.M.A.K., and K.B.T.; funding acquisition: S.K.C., O.J.L., A.H.S., M.M.A.K., K.P.L., M.M., and M.M.

Funding: This research was funded by Universiti Putra Malaysia through the Putra-Grant (vote no.: 9552300). N. M. Hapipi would like to acknowledge financial support from the Ministry of Education Malaysia through the MyMaster scholarship, Universiti Putra Malaysia under the Graduate Research Fellowship (GRF). The authors are also grateful to Japan Science and Technology Agency for their financial assistance through SAKURA Exchange Program in Science under Shibaura Institute of Technology.

Acknowledgments: We would like to thank Nik Afida Anis Azahari, Kamsiah Alias and Norhaslinda Noruddin for their technical support.

Conflicts of Interest: The authors declare no conflict of interest.

References

1. Barnes, P.N.; Haugan, T.J.; Baca, F.J.; Varanasi, C.V.; Wheeler, R.; Meisenkothen, F.; Sathiraju, S. Inducing self-assembly of Y_2BaCuO_5 nanoparticles via Ca-doping for improved pinning in $YBa_2Cu_3O_{7-x}$. *Physica C* **2009**, *469*, 2029–2032. [CrossRef]

2. Horvath, D.; Harnois, C.; Noudem, J.G. Li and Ce doping of melt-textured YBCO: Improved J_c at medium fields. *Mater. Sci. Eng. B* **2008**, *151*, 36–39. [CrossRef]

3. Klie, R.F.; Buban, J.P.; Varela, M.; Franceschetti, A.; Jooss, C.; Zhu, Y.; Browning, N.D.; Pantelides, S.T.; Pennycook, S.J. Enhanced current transport at grain boundaries in high-T_c superconductors. *Nature* **2005**, *435*, 475–478. [CrossRef]

4. Jin, L.H.; Zhang, S.N.; Yu, Z.M.; Li, C.S.; Feng, J.Q.; Sulpice, A.; Wang, Y.; Zhang, P.X. Influences of $BaZrO_3$ particles on the microstructure and flux pinning of YBCO film prepared by using modified TFA-MOD approach. *Mater. Chem. Phys.* **2015**, *149–150*, 188–192. [CrossRef]

5. Awano, M.; Fujishiro, Y.; Moon, J.; Takagi, H.; Rybchenko, S.; Bredikhin, S. Microstructure control of an oxide superconductor on interaction of pinning centers and growing crystal surface. *Physica C* **2000**, *341–348*, 2017–2018. [CrossRef]

6. Paulose, K.V.; Koshy, J.; Damodaran, A.D. Superconductivity in $YBa_2Cu_3O_{7-\delta}$-ZrO_2 systems. *Supercond. Sci. Technol.* **1991**, *4*, 96–101. [CrossRef]

7. Palonen, H.; Huhtinen, H.; Shakhov, M.A.; Paturi, P. Electron mass anisotropy of $BaZrO_3$ doped YBCO thin films in pulsed magnetic fields up to 30 T. *Supercond. Sci. Technol.* **2013**, *26*, 045003. [CrossRef]

8. Bretos, I.; Schneller, T.; Falter, M.; Backer, M.; Hollmann, E.; Wordenweber, R.; Molina-Luna, L.; Tendelooe, G.V.; Eibl, O. Solution-derived $YBa_2Cu_3O_{7-\delta}$ (YBCO) superconducting films with $BaZrO_3$ (BZO) nanodots based on reverse micelle stabilized nanoparticles. *J. Mater. Chem. C* **2015**, *3*, 3971–3979. [CrossRef]

9. Peurla, M.; Paturi, P.; Stepanov, Y.P.; Huhtinen, H.; Tse, Y.Y.; Bodi, A.C.; Raittila, J.; Laiho, R. Optimization of the $BaZrO_3$ concentration in YBCO films prepared by pulsed laser deposition. *Supercond. Sci. Technol.* **2006**, *19*, 767–771. [CrossRef]

10. Malmivirta, M.; Rijckaert, H.; Paasonen, V.; Huhtinen, H.; Hynninen, T.; Jha, R.; Awana, V.S.; Driessche, I.V.; Paturi, P. Enhanced flux pinning in YBCO multilayer films with BCO nanodots and segmented BZO nanorods. *Sci. Rep.* **2017**, *7*, 14682. [CrossRef]

11. Wang, F.; Tian, H. $BaZrO_3$ (BZO) nanoparticles as effective pinning centers for -$YBa_2Cu_3O_{7-\delta}$ (YBCO) superconducting thin films. *J. Mater. Sci. Mater. Electron.* **2019**, *30*, 4137–4143. [CrossRef]

12. Jha, A.K.; Khare, N. Investigation of flux pinning properties of YBCO: $BaZrO_3$ composite superconductor from temperature dependent magnetization studies. *J. Magn. Magn. Mater.* **2010**, *322*, 2653–2657. [CrossRef]

13. Arlina, A.; Halim, S.A.; Awang Kechik, M.M.; Chen, S.K. Superconductivity in Bi-Pb-Sr-Ca-Cu-O ceramics with YBCO as additive. *J. Alloys Compd.* **2015**, *645*, 269–273. [CrossRef]

14. Paz-Pujalt, G.R.; Mehrotra, A.K.; Ferranti, S.A.; Agostinelli, J.A. Solid state reactions in the formation of $YBa_2Cu_3O_{7-\delta}$ high T_c superconductor powders. *Solid State Ionics* **1989**, *32–33*, 1179–1182. [CrossRef]

15. Fujihara, S.; Kozuka, H.; Yoko, T.; Sakka, S. Mechanism of formation of $YBa_2Cu_4O_8$ superconductor in the sol-gel synthesis. *J. Sol-Gel Sci. Technol.* **1994**, *1*, 133–140. [CrossRef]

16. Ramli, A.; Halim, S.A.; Chen, S.K.; Awang Kechik, M.M. The effect of Gd_2O_3 nanoparticles addition on microstructural and electrical properties of YBCO superconductor. *ARPN J. Eng. Appl. Sci.* **2016**, *11*, 13708–13715.

17. Hamadneh, I.; Rosli, A.M.; Abd-Shukor, R.; Suib, N.R.M.; Yahya, S.Y. Superconductivity of $REBa_2Cu_3O_{7-\delta}$ (RE = Y, Dy, Er) ceramic synthesized via coprecipitation method. *J. Phys. Conf. Ser.* **2008**, *97*, 012063. [CrossRef]

18. Bhargava, A.; Mackinnon, I.D.R.; Yamashita, T.; Page, D. Bulk manufacture of YBCO powders by coprecipitation. *Physica C* **1995**, *241*, 53–62. [CrossRef]

19. Mousa Dihom, M.; Shaari, A.H.; Baqiah, H.; Al-Hada, N.M.; Talib, Z.A.; Chen, S.K.; Aziz, R.S.; Awang Kechik, M.M.; Lim, K.P.; Shukor, R.A. Structural and superconducting properties of $Y(Ba_{1-x}K_x)_2Cu_3O_{7-\delta}$ ceramics. *Ceram. Int.* **2017**, *43*, 11339–11344. [CrossRef]

20. Wahid, M.H.; Zainal, Z.; Hamadneh, I.; Tan, K.B.; Halim, S.A.; Rosli, A.M.; Alaghbari, E.S.; Nazarudin, M.F.; Kadri, E.F. Phase formation of $REBa_2Cu_3O_{7-\delta}$ (RE: $Y_{0.5}Gd_{0.5}$, $Y_{0.5}Nd_{0.5}$, $Nd_{0.5}Gd_{0.5}$) superconductors from nanopowders synthesised via co-precipitation. *Ceram. Int.* **2012**, *38*, 1187–1193. [CrossRef]

21. Ochsenkuhn-Petropoulou, M.; Argyropoulou, R.; Tarantilis, P.; Kokkinos, E.; Ochsenkuhn, K.M.; Parissakis, G. Comparison of the oxalate co-precipitation and the solid state reaction methods for the production of high temperature superconducting powders and coating. *J. Mater. Process. Technol.* **2002**, *127*, 122–128. [CrossRef]

22. Mohd Hapipi, N.; Chen, S.K.; Shaari, A.H.; Awang Kechik, M.M.; Tan, K.B.; Lim, K.P. Superconductivity of Y_2O_3 and $BaZrO_3$ nanoparticles co-added $YBa_2Cu_3O_{7-\delta}$ bulks prepared using co-precipitation method. *J. Mater. Sci. Mater. Electron.* **2018**, *29*, 18684–18692. [CrossRef]

23. Mohd Hapipi, N.; Chen, S.K.; Shaari, A.H.; Awang Kechik, M.M.; Tan, K.B.; Lim, K.P.; Lee, O.J. AC susceptibility of $BaZrO_3$ nanoparticles added $YBa_2Cu_3O_{7-\delta}$ superconductor prepared via coprecipitation method. *J. Supercond. Novel Magn.* **2018**, *32*, 1191–1198. [CrossRef]

24. Patnaik, P. *Dean's Analytical Chemistry Handbook*, 2nd ed.; McGraw Hill Professional: New York, NY, USA, 2004.

25. Zhang, C.J.; Oyanagi, H. The synthesis condition and its influence on T_c in Mn doped $La_{1.85}Sr_{0.15}CuO_4$. *Physica C* **2008**, *468*, 1155–1158. [CrossRef]

26. Pomar, A.; Vlad, V.R.; Llordes, A.; Palau, A.; Gutiérrez, J.; Ricart, S.; Puig, T.; Obradors, X.; Usoskin, A. Enhanced vortex pinning in YBCO coated conductors with BZO nanoparticles from chemical solution deposition. *IEEE Trans. Appl. Supercond.* **2009**, *19*, 3258–3261. [CrossRef]

27. Cracolice, M.S.; Peters, E.I. *Introductory Chemistry: An Active Learning Approach*; Brooks Cole: Belmont, CA, USA, 2012.

28. Langford, J.I.; Wilson, A.J.C. Scherrer after sixty years: A survey and some new results in the determination of crystallite size. *J. Appl. Crystallogr.* **1978**, *11*, 102–113. [CrossRef]

29. Hassanzadeh-Tabrizi, S.A.; Mazaheri, M.; Aminzare, M.; Sadrnezhaad, S.K. Reverse precipitation synthesis and characterization of CeO_2 nanopowder. *J. Alloys Compd.* **2010**, *491*, 499–502. [CrossRef]

30. Mondal, O.; Pal, M.; Singh, R.; Sen, D.; Mazumder, S.; Pal, M. Influence of doping on crystal growth, structure and optical properties of nanocrystalline $CaTiO_3$: A case study using small-angle neutron scattering. *J. Appl. Crystallogr.* **2015**, *48*, 836–843. [CrossRef]

31. Luo, Y.Y.; Wu, Y.C.; Xiong, X.M.; Li, Q.Y.; Gawalek, W.; He, Z.H. Effects of precursors with fine $BaZrO_3$ inclusions on the growth and microstructure of textured YBCO. *J. Supercond.* **2000**, *13*, 575–581. [CrossRef]

32. Nikolo, M. Superconductivity: A guide to alternating current susceptibility measurements and alternating current susceptometer design. *Am. J. Phys.* **1995**, *63*, 57–65. [CrossRef]

33. Ciontea, L.; Celentano, G.; Augieri, A.; Ristoiu, T.; Suciu, R.; Gabor, M.S.; Rufoloni, A.; Vannozzi, A.; Galluzi, V.; Petrisor, T. Chemically processed $BaZrO_3$ nanopowders as artificial pinning centres. *J. Phys. Conf. Ser.* **2008**, *97*, 012289. [CrossRef]

34. MacManus-Driscoll, J.L.; Foltyn, S.R.; Jia, Q.X.; Wang, H.; Serquis, A.; Civale, L.; Maiorov, B.; Hawley, M.E.; Maley, M.P.; Peterson, D.E. Strongly enhanced current densities in superconducting coated conductors of $YBa_2Cu_3O_{7-x}$ + $BaZrO_3$. *Nat. Mater.* **2004**, *3*, 439–443. [CrossRef] [PubMed]

35. Kameli, P.; Salamati, H.; Eslami, M. The effect of sintering temperature on the intergranular properties of Bi2223 superconductors. *Solid State Commun.* **2006**, *137*, 30–35. [CrossRef]

36. Nedkov, I.; Veneva, A. Grain boundaries contribution to the complex susceptibility of KCu-doped YBCO high-temperature superconductors. *J. Low Temp. Phys.* **1997**, *107*, 497–502. [CrossRef]

37. Deac, I.G.; Burzo, E.; Pop, A.V.; Pop, V.; Tetean, R.; Kovacs, D.; Borodi, G. Intergranular properties of $(Y_{1-x-y}Zr_xCa_y)Ba_2Cu_3O_{7-\delta}$ compounds. *Int. J. Mod. Phys. B* **1999**, *13*, 1645–1654. [CrossRef]

38. Sbarciog, C.; Redac, R.T.; Deac, I.G.; Pop, I. Intergranular properties of Zr-substituted Y123 compounds. *Mod. Phys. Lett. B* **2006**, *20*, 1191–1198. [CrossRef]

39. Rani, P.; Jha, R.; Awana, V.P.S. AC susceptibility study of superconducting $YBa_2Cu_3O_7$: Ag_x bulk composites (x = 0.0–0.20): The role of intra and intergranular coupling. *J. Supercond. Novel Magn.* **2013**, *26*, 2347–2352. [CrossRef]

40. Nur-Akasyah, J.; Nur-Shamimie, N.H.; Abd-Shukor, R. Effect of CdTe addition on the electrical properties and AC susceptibility of $YBa_2Cu_3O_{7-\delta}$ superconductor. *J. Supercond. Novel Magn.* **2017**, *30*, 3361–3365. [CrossRef]

41. Clem, J.R. Granular and superconducting-glass properties of the high-temperature superconductors. *Physica C* **1988**, *153–155*, 50–55. [CrossRef]

42. Emmen, J.H.P.M.; Brabers, V.A.M.; De Jonge, W.J.M.; Steen, C.V.D.; Dalderop, J.H.J.; Geppaart, P.M.A.; Kopinga, K. Microstructure and properties of Bi-Ca-Sr-Cu-O superconductors. *J. Less-Common Met.* **1989**, *151*, 63–69. [CrossRef]

43. Narlikar, A.V. *Field Penetration and Magnetization of High Temperature Superconductors*; Nova Science Publisher, Inc.: New York, NY, USA, 1995.

© 2019 by the authors. Licensee MDPI, Basel, Switzerland. This article is an open access article distributed under the terms and conditions of the Creative Commons Attribution (CC BY) license (http://creativecommons.org/licenses/by/4.0/).

Article

Two- and Three-Dimensional Superconducting Phases in the Weyl Semimetal TaP at Ambient Pressure

Maarten R. van Delft [1,2,†], **Sergio Pezzini** [1,2,‡], **Markus König** [3], **Paul Tinnemans** [2], **Nigel E. Hussey** [1,2] and **Steffen Wiedmann** [1,2,*]

1   High Field Magnet Laboratory (HFML-EMFL), Radboud University, Toernooiveld 7, 6525 ED Nijmegen, The Netherlands; maarten.vandelft@epfl.ch
2   Radboud University, Institute for Molecules and Materials, 6525 AJ Nijmegen, The Netherlands
3   Max Planck Institute for Chemical Physics of Solids, Nöthnitzer Straße 40, D-01187 Dresden, Germany
*   Correspondence: Steffen.Wiedmann@ru.nl
†   Current address: Laboratory of Quantum Materials (QMAT), Institute of Materials (IMX), École Polytechnique Fédérale de Lausanne (EPFL), 1015 Lausanne, Switzerland.
‡   Current address: Center for Nanotechnology Innovation @NEST-Istituto Italiano di Tecnologia, Piazza San Silvestro 12, 56127 Pisa, Italy.

Received: 5 March 2020; Accepted: 4 April 2020; Published: 10 April 2020

Abstract: The motivation to search for signatures of superconductivity in Weyl semi-metals and other topological phases lies in their potential for hosting exotic phenomena such as nonzero-momentum pairing or the Majorana fermion, a viable candidate for the ultimate realization of a scalable quantum computer. Until now, however, all known reports of superconductivity in type-I Weyl semi-metals have arisen through surface contact with a sharp tip, focused ion-beam surface treatment or the application of high pressures. Here, we demonstrate the observation of superconductivity in single crystals, even an as-grown crystal, of the Weyl semi-metal tantalum phosphide (TaP), at ambient pressure. A superconducting transition temperature, T_c, varying between 1.7 and 5.3 K, is observed in different samples, both as-grown and microscopic samples processed with focused ion beam (FIB) etching. Our data show that the superconductivity present in the as-grown crystal is inhomogeneous yet three-dimensional. For samples fabricated with FIB, we observe, in addition to the three-dimensional superconductivity, a second superconducting phase that resides on the sample surface. Through measurements of the characteristic fields as a function of temperature and angle, we are able to confirm three dimensionality of the two distinct superconducting phases.

Keywords: superconductivity; Weyl semimetal; focused ion beam

---

1. Introduction

Since the discovery of Weyl semimetals, a great deal of work has been devoted to understanding the properties of these topological materials, whose band structure includes specific points known as Weyl nodes where non-degenerate bands touch each other and disperse linearly. Weyl semimetals differ from the related Dirac semimetals in that they require either time-reversal or inversion symmetry to be broken in order to lift the degeneracy of the nodes. Consequently, Weyl nodes always exist in pairs of opposite chirality that are connected through Fermi arcs running along the surface of the material, as has been observed experimentally in several materials using angle-resolved photoemission spectroscopy (ARPES) [1–7]. The chiral nature of the Weyl nodes can furthermore manifest itself in electrical transport, in the form of the chiral anomaly, leading to a negative longitudinal magnetoresistance. The observation of this effect has been reported in several materials [8–12], but its origin remains uncertain [13–15].

In addition to the intrinsic transport properties of Weyl semimetals, the combination of Weyl physics and superconductivity may support Majorana [16] or other exotic surface states [17] as a result of their topological nature. These states are of fundamental interest, as ultimately, they may be applicable in the field of quantum computation. For this reason, there is an ongoing effort aimed at achieving superconductivity in such materials and investigating their properties, either through the use of the proximity effect [16,18] or by other means. Successes in this area have been achieved for the type-II Weyl semimetals $MoTe_{2-x}S_x$ [19] and $TaIrTe_4$ [20], which exhibit intrinsic, exotic superconductivity. For type-I Weyl semimetals, however, such observations are still lacking.

The family of compounds comprising TaP, TaAs, NbAs and NbP are all experimentally confirmed as type-I Weyl semimetals [2,3,5,21–23] and are now among the most ardently studied compounds in this class. Under the right conditions, each member of the family has shown trace signatures of superconductivity. In both TaAs and TaP, a superconducting onset has been induced through contact with sharp tips of different materials [24–27], while in TaP alone, this onset was also achieved under the application of extremely high pressures [28]. In the former case, however, the specific technique does not allow for the observation of zero resistance, while in the latter, only a partial drop in the resistivity was observed. Similarly, signs of partial superconductivity were also seen in polycrystalline [29] or powder [30] samples of NbP. A state of zero resistance, however, has only been reported in samples where superconductivity was induced in a thin surface layer by treatment with focused ion beam (FIB) [31]. To date, no trace of superconductivity at ambient pressure has been reported in pristine single crystals of any member of this family.

In this work, we demonstrate the onset of intrinsic three-dimensional superconductivity in TaP, in addition to the FIB-induced surface superconductivity that was previously reported by Bachmann et al. [31]. In one of our FIB-processed crystals, both two- and three-dimensional superconductivity are observed, while in the as-grown parent crystal only three-dimensional but inhomogeneous (filamentary) superconductivity appears. In both cases, we find T_c to vary between 1.7 and 5.3 K. Finally, we delineate and identify the two transitions in the microfabricated crystal based on the angle dependence of its characteristic magnetic fields. The presence of three-dimensional superconductivity in the microfabricated crystal, as revealed through this analysis, confirms that superconductivity develops within the bulk of the as-grown crystal. Though its origin could not be confirmed in this initial study, we speculate here that it may arise from domains within the crystal with excess Ta.

2. Results

Figure 1b shows low-temperature resistivity curves for both the pristine crystal and one of the microstructured samples, labelled sample 4 (data for all four microstructured samples can be found in Figure S1 of the Supplementary Information). The parent crystal exhibits an incomplete, resistive transition at a transition temperature $T_c = 3 \pm 0.8$ K that resembles the one which had been observed previously in TaP under high pressure [28]. The microstructured crystal, by contrast, exhibits a sharp transition to zero resistance with $T_c = 3.3 \pm 0.5$ K. Note that the other microstructured crystals showed either broad and/or incomplete transitions. As a consequence, we focussed our subsequent measurements and analysis on sample 4.

In order to confirm the presence of superconductivity in our samples, we studied the evolution of the resistive transition in the presence of a magnetic field. Figure 1c shows temperature sweeps of the resistivity of one of the microstructured samples under different, constant magnetic fields, applied perpendicular to the current direction. Clearly, the resistive transition in the FIB-processed sample is suppressed gradually with field, as is expected in the case of a superconducting transition. For $B_\perp = 10$ T, it vanishes completely. In the parent crystal, the transition is suppressed more strongly and vanishes under a very small magnetic field.

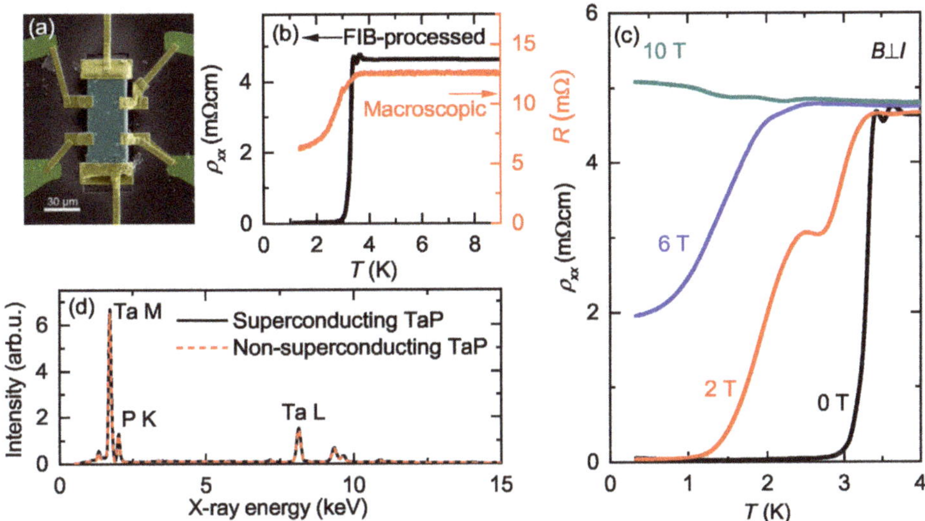

Figure 1. (**a**) False color SEM image of a typical FIB contacted crystal. (**b**) Temperature sweeps showing the resistive transition of the FIB-processed sample 4 and the macroscopic crystal from which all FIB samples were cut. (**c**) Resistivity curves at low temperature of FIB-processed sample 4, under different, constant magnetic fields perpendicular to the current direction. (**d**) EDX spectra of the bulk sample used in our experiments and a non-superconducting sample obtained from a different source. Within the resolution, the compositions are the same.

Superconductivity has been reported before in microfabricated samples of TaP, TaAs, NbP and NbAs [31], but in as-grown TaP it has never been observed at ambient pressure despite the large number of measurements at low temperature that have been carried out on this material [32–35]. For this reason, we investigated our bulk crystal for any deviations in terms of composition and structure that might indicate the origin of the superconductivity.

We studied the parent crystal with EDX and compared the stoichiometry with that of a non-superconducting crystal of TaP from a different source. The resulting spectra are shown in Figure 1d. From these spectra, we find that the two crystals have a stoichiometry that is identical to within an experimental error of about 1%, implying that our material has a very similar composition to that used in other studies. Furthermore, XRD measurements (see Supplementary Information, Figure S8) confirm that our TaP crystal is in the $I4_1md$ space group, as is usual for TaP under ambient conditions [21,36,37]. This structure is different, however, to that found in TaP under high pressure or in MoP (also under pressure), where in both cases, superconductivity appears in the P-$6m2$ phase. Thus the superconducting state that develops in our crystals appears to be distinct from what has been observed before.

Figure 2a shows field sweeps of the resistivity of one of the microstructured crystals (sample 4) under different orientations of the magnetic field. Here, the angle 90° denotes a field parallel to the direction of the applied current (see inset of Figure 2b), which in this case is the crystallographic *ab*-plane. For angles away from 90°, there is a clear double transition that we associate with distinct superconducting phases. Close to 90°, the characteristic fields of the two superconducting phases appear to merge, causing the two transitions to become indistinguishable.

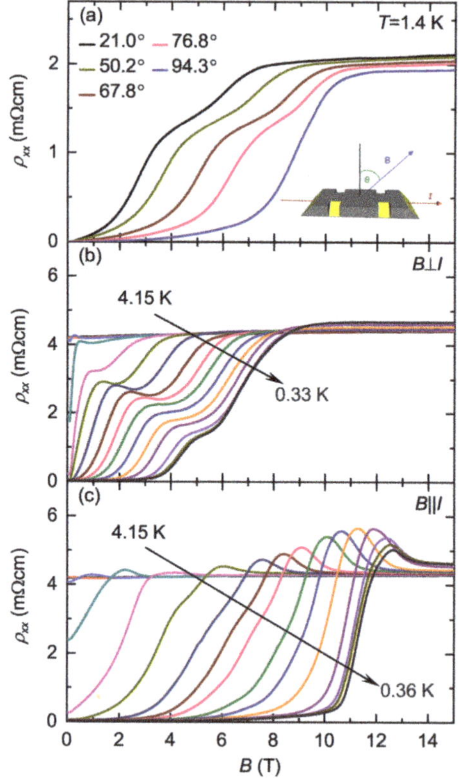

Figure 2. Field sweeps of the resistivity of the FIB-processed sample 4 for (**a**) different angles as defined in the inset and (**b**) different temperatures in a magnetic field parallel to the current direction and (**c**) in a field perpendicular to the current.

Figure 2b,c show the temperature dependence of the resistive transitions in field, for fields perpendicular and parallel to the current direction, respectively. In perpendicular field (Figure 2b), there are two distinct superconducting features separated by a broad shoulder. In the parallel field configuration on the other hand, the shoulder is much weaker. This suggests that the characteristic fields of the two states are of similar magnitude, but have a different temperature dependence. The data also reveal an upturn in the resistivity before the normal state is fully restored. Such an upturn is frequently observed in inhomogeneous superconductors [38–42] due to current redistribution as some parts of the crystal turn superconducting while others remain resistive [43]. It was also recently seen in the pressure-induced superconducting state of MoP [44].

3. Discussion

3.1. Dimensionality of the Superconducting Phases

The data in Figure 2 can be used to extract characteristic fields for both transitions as a function of angle and temperature, the results of which are shown in Figure 3. The two field scales are determined by defining the field at which the resistance has risen to 90% of the normal state (H_2) and the field that minimizes the second derivative of the resistivity (H_1) (see Supplementary Figure S2 for more detail). In Figure 3a, we show both H_1 and H_2 as a function of the angle between the magnetic field and the *c*-axis. The solid and dashed lines are fits to the two-dimensional Tinkham model [45] and the three-dimensional Ginzburg–Landau (GL) model [46], respectively. It is found that the 2D Tinkham model gives an excellent description of the behavior of H_1, covering not only the cusp at 90° that the 3D GL model misses, but also its behavior in near perpendicular fields. Conversely, the angle dependence of H_2 does not have a sharp cusp like H_1 and is better described by the 3D GL model.

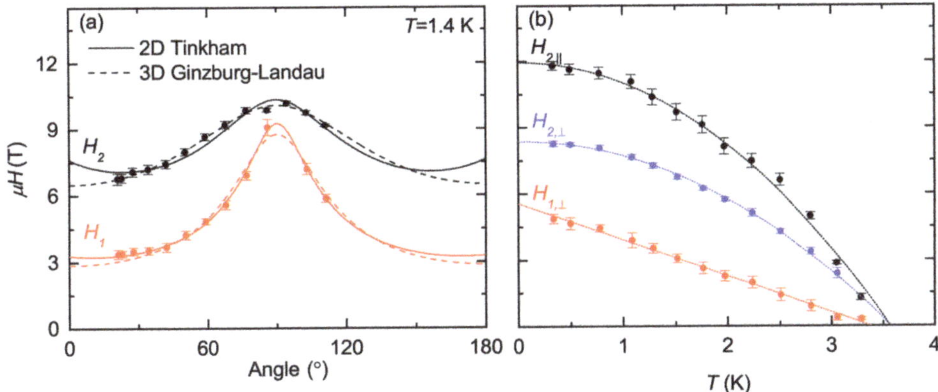

Figure 3. (**a**) Angle dependence of the characteristic fields, fitted with the two-dimensional Tinkham model (solid lines) and the three-dimensional Ginzburg–Landau model (dashed lines). (**b**) Temperature dependence of the characteristic fields for parallel and perpendicular fields. H_2 is fitted using a three-dimensional model and H_1 with the two-dimensional Ginzburg–Landau model. H_1 in parallel field cannot be unambiguously identified and is thus omitted from this figure.

Some refinements of the 2D Tinkham model are possible (outlined in the Supplementary Information), one for the case of intrinsic surface superconductivity and another for a thin superconducting film with a thickness less than or comparable to the coherence length ξ_{GL}, a less stringent requirement than that of the Tinkham model which requires a thickness $d \ll \xi_{GL}$ [47,48]. Both of these adapted models, however, lead to a less accurate fitting for both H_1 and H_2 (see Supplementary Figure S3). Thus, with the GL and Tinkham models giving the most accurate descriptions of H_2 and H_1 respectively, we conclude that H_2 arises from a certain superconducting volume fraction within the interior of the crystal while H_1 is characteristic of a 2D superconducting state, presumably arising from a very thin layer on the surface that satisfies the criterion of $d \ll \xi_{GL}$.

Figure 3b shows the phase diagram of FIB-processed TaP, in both parallel and perpendicular field configurations. H_1 is only shown for perpendicular field, as it is not distinguishable in the parallel field configuration. The behavior of H_2 in both configurations can be well described by the expression $H_2 = H_2(0)(1 - (T/T_c)^2)$, in agreement with the description of the critical field for a three-dimensional GL superconductor. H_1 on the other hand, is better described with a linear temperature dependence, consistent with the GL model for two-dimensional superconductors: $\mu H_{1,\perp} = \Phi_0/(2\pi\xi_{GL}^2)(1 - T/T_c)$ [49–51]. Associating H_1 with the upper critical field for the two-dimensional superconducting layer, we can estimate ξ_{GL} to be approximately 7.7 nm.

The origin of the 2D superconductivity is believed to be the presence of excess Ta induced during FIB milling. In order to make an estimate of the thickness d of the superconducting layer, we use the SRIM-2013 [52] code to simulate the ion milling process. With the low acceleration voltage of 8 kV that is used for the final polishing of the sample, the Ga^+ ions penetrate about 5.0 nm below the surface of the sample. As P is sputtered approximately two times more readily than Ta, an average composition is expected in this thin layer of $Ta_{2.1}P$. Of course, the thickness of the superconducting layer cannot simply be assumed to be the same as the ion penetration depth; if anything, this depth gives an upper limit. Considering this, it is not unreasonable to expect that the inequality $d \ll \xi_{GL}$ is indeed satisfied.

3.2. Critical Currents

Information on the evolution of the two superconducting phases in sample 4 with temperature can be gleaned by looking at the critical currents associated with the transitions. To this end, we measured the current-voltage (IV) characteristics and differential resistivity curves in different magnetic fields and temperatures. In Figure 4, the data are shown for zero field at $T = 1.3$ K (see Supplementary Figures S4 and S5 for the full set of data in different magnetic fields and at different temperatures).

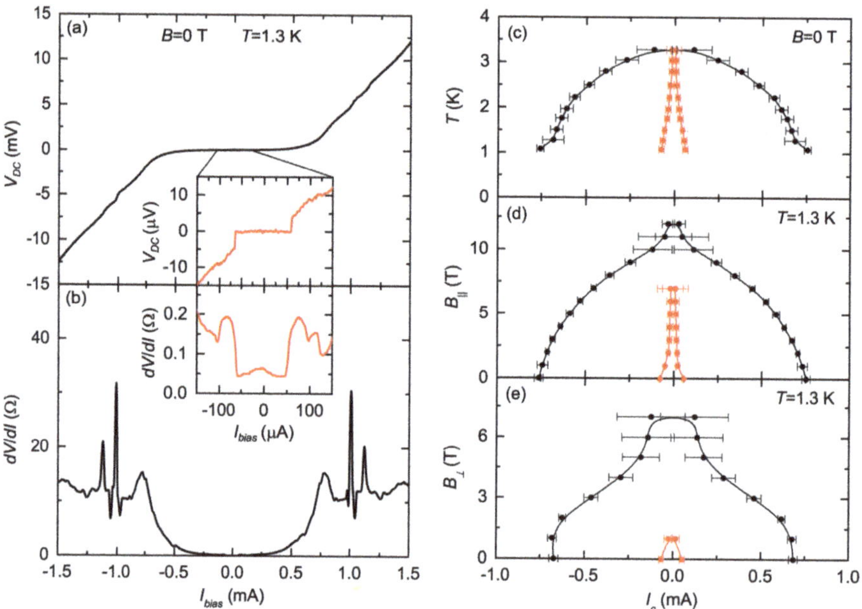

Figure 4. (**a**) IV curve of sample 4 measured at $T = 1.3$ K and $B = 0$ T. Inset: expanded view on the low-current region where the transition due to the surface can be seen. (**b**) Differential resistance measured simultaneously with the IV curve. Inset: differential resistance corresponding to the inset of (**a**). (**c**–**e**) Critical currents of the superconducting volume (black) and surface (red) as a function of (**c**) the temperature, (**d**) parallel magnetic field and (**e**) perpendicular magnetic field.

Several clear transitions can be seen in the differential curves shown in Figure 4b. Around 1.0 and 1.1 mA, there are two distinct features that do not lead to a significant change in resistance. We presume that these correspond to parts of the bulk crystal becoming superconducting while others remain resistive and are likely to be a consequence of strong inhomogeneity present in the sample. The transition at 0.7 mA then represents the majority of the bulk crystal becoming normal, leading to a strong increase in the resistivity. The state of zero resistance, however, can only be seen below a much smaller bias current of about 60 µA, as seen in the insets of Figure 4a,b. Considering the two-dimensional nature of the superconductivity within the surface layer, this small bias current corresponds to a current density of approximately 8×10^4 A·cm^{-2}, much larger than that associated with the three-dimensional transition (about 1×10^3 A·cm^{-2}).

Figure 4c–e shows the transition temperatures and magnetic fields associated with the two- and three-dimensional superconducting phases of the FIB-processed TaP as a function of the critical currents. This is a further confirmation of Figure 3 as the two transitions are suppressed at the same temperature, but at different values of the magnetic field. In perpendicular field, the feature associated with surface superconductivity can be seen up to 1 T, whereas in parallel field it survives up to 7 T. The difference is less apparent for the three-dimensional superconductivity, but also this can be seen to persist to higher fields in the parallel configuration, in agreement with Figure 3b.

3.3. BCS–BEC Crossover Regime

Via Hall effect measurements, the carrier concentration n can be estimated for each of the microstructured samples. As shown in Figure S6a, the Hall resistivity $\rho_{xy}(H)$ of sample 4 was found to be non-linear, as expected for a semi-metal in which the densities and/or mobilities of the electron- and hole-like carriers are distinct. From the initial slope of $\rho_{xy}(H)$, we can nevertheless obtain a maximum value for the density of a single carrier type for each sample. These range from 4.5×10^{18} to 3.9×10^{19} cm^{-3}. For a semimetal such as TaP, these are typical values, in agreement with the literature [32,34]. However, for a superconducting material, these are unusually low carrier concentrations. For comparison, we consider SrTiO$_3$, for which a carrier concentration of 4.1×10^{18} with a T_c of 180 mK has been reported [53]. The T_c of our TaP is at least 10 times higher despite the fact that the carrier densities are comparable. Additionally, the observed characteristic critical fields are relatively high, exceeding the usual Pauli paramagnetic limit. For these reasons, we consider the possibility of a crossover between a Bardeen–Cooper–Schrieffer state and a Bose–Einstein condensate (BCS–BEC) for the observed superconductivity.

Using the value of 7.7 nm for the coherence length ξ obtained from the fitting of Figure 3, we calculate the number of pairs within a coherence volume $V_{coh} = 4/3\pi\xi^3$. In sample 4, this amounts to approximately 25 pairs, suggesting there is limited overlap between the pairs. Typically, a BCS superconductor has many thousands of pairs overlapping within V_{coh}, whereas a BEC superconductor has less than one pair in V_{coh} and there is no interaction between different pairs. With a number of 25 pairs in V_{coh}, sample 4 is similar to FeSe (with 31 pairs [54]), which is considered to reside within the BCS–BEC crossover regime [55–57]. These findings therefore suggest that this new breed of semi-metals is a good playground for the observation of possible exotic superconductivity on the BCS–BEC boundary.

4. Conclusions

In conclusion, we have established the existence of three-dimensional, inhomogeneous superconductivity at ambient pressure in a crystal of TaP and confirmed the appearance of FIB-induced two-dimensional surface superconductivity. In other studies, it was found that TaP typically contains a large density of defects and can be off-stoichiometric with an excess of Ta [33,36]. Our EDX data do not exclude off-stoichiometry in our samples, but considering the similarity between the superconducting and non-superconducting samples, any overall off-stoichiometry cannot explain the superconductivity we observe.

It is, however, apparent that our sample is strongly inhomogeneous as we see multiple critical currents associated with partial superconducting transitions, as well as an upturn of the resistance just above the critical field in the parallel field configuration. As such, there may exist domains with a local excess of Ta or defect structures that support superconductivity. Nevertheless, our findings call for a thorough study of the growth of TaP and related compounds in order to establish under what conditions superconductivity can be optimized. Further research is also required to ascertain whether the superconducting volume in TaP retains all the characteristics of a Weyl semimetal. We note here that none of our superconducting crystals exhibited a negative longitudinal magnetoresistance, for example. If it does, however, TaP may provide an ideal platform for the study of Weyl superconductivity.

5. Materials and Methods

5.1. Sample Preparation

The single crystal used in this study was grown by chemical vapor transport using polycrystalline TaP as a source material [37]. Via X-ray crystallography, a facet matching the ab-plane of this crystal was identified and the microsamples were cut from this facet using focused ion beam (FIB) milling. For the rough cutting, an acceleration voltage of 30 kV with a large current of 20 nA was used. Initially, a rectangular piece of the crystal was cut out and remained attached only via a thin bar to the main crystal. A micromanipulator was then brought in contact with the rectangle and fixed onto it via Pt deposition, after which the bar was cut through and the sample was transferred to a silicon oxide substrate with prepatterned gold contact pads. On the substrate, a further shaping of the sample took place at 30 kV and 0.9 nA, followed by a more precise cleaning at 8 kV and 0.2 nA. Contacts were made between the sample and the gold pads via FIB-induced Pt deposition in a standard Hall-bar configuration (see Figure 1b).

5.2. Resistivity Measurements

The resistivity measurements were performed in either a superconducting magnet with a maximum field of 15 T, using a ^{3}He cryostat, or a resistive magnet of 33 T with a ^{4}He cryostat with base temperature 1.3 K. We mounted the samples on a rotatable platform in order to vary their angle with the field. Each microsample had six contacts, with the two covering the short ends of the sample being used as current contacts and the other contacts to measure either a longitudinal or a hall voltage. We used an AC current excitation of 10 or 100 μA and acquisition took place with standard lock-in techniques. For the measurement of differential resistance, we used a voltage source in series with a 100 kΩ resistor to supply the DC bias current, with the lock-in amplifier similarly supplying a 10 μA AC current on top. A multimeter was placed in parallel with the lock-in amplifier to measure the DC signal.

The as-grown sample had an arbitrary shape and so it could not be contacted in any well-defined geometry, making it impossible to determine the absolute resistivity. This sample was measured with four contacts placed along the crystal and a current of 0.5 mA.

5.3. Energy-Dispersive X-ray Analysis (EDX)

In order to determine whether off-stoichiometry might be responsible for the observed superconductivity in our samples, we performed an elemental analysis of the parent crystal together with a non-superconducting crystal of TaP from a different source. The two bulk crystals were measured in the same EDX system during a single run to exclude any difference in signal other than from the material itself. We aligned the crystals by eye to have a flat surface facing the electron beam. To correct for any small discrepancies remaining in the angle between the electron beam and the crystal, both crystals were measured twice with a 180° in-plane rotation in between and the two measurements were averaged. The results before and after the rotation were comparable, suggesting that the orientation and flatness of the surfaces were good.

Despite our best efforts, we could not determine the precise Ta:P ratio with real confidence due to the inherent difficulties associated with quantitative EDX measurements. Such a measurement would require a reference TaP sample that is precisely stoichiometric. While our reference sample is known to be non-superconducting, its stoichiometry is not guaranteed.

5.4. X-ray Diffraction (XRD)

Reflections were measured on a Bruker D8 Quest diffractometer with sealed tube and Triumph monochromator ($\lambda = 0.71073$ Å). The unit cell was found using the software CELL_NOW [58]. The XRD pattern is presented in supplementary Figure S8.

Supplementary Materials: The following are available online at http://www.mdpi.com/2073-4352/10/4/288/s1: Section I: Estimation of uncertainty. Section II: Fitting models. Section III with Table S1: list of sample dimensions. Figure S1: Temperature and field sweeps of all crystals. Figure S2: Illustration of characteristic field identification. Figure S3: Angle dependence of characteristic fields with different fittings. Figure S4: IV curves and dV/dI of sample 4. Figure S5: IV curves and dV/dI of sample 4, highlighting the low-current region. Figure S6: Quantum oscillation data of sample 4. Figure S7: Magnetoresistance data of sample 1. Figure S8: Powder XRD spectrum of the parent crystal.

Author Contributions: Conceptualization, M.R.v.D. and S.W.; methodology, M.R.v.D.; validation, M.R.v.D., S.P., N.E.H. and S.W.; formal analysis, M.R.v.D.; investigation, M.R.v.D., S.P., M.K. and P.T.; resources, S.W.; data curation, M.R.v.D.; writing—original draft preparation, M.R.v.D.; writing—review and editing, M.R.v.D., S.P., N.E.H. and S.W.; visualization, M.R.v.D.; supervision, S.W. and N.E.H.; project administration, S.W. and N.E.H.; funding acquisition, S.W. and N.E.H. All authors have read and agreed to the published version of the manuscript.

Funding: This research was supported by the HFML-RU/FOM, member of the European Magnetic Field Laboratory (EMFL).

Acknowledgments: The authors would like to thank Nan Xu for providing the sample used to conduct this study and Andrew Mackenzie for help with the interpretation of the EDX data.

Conflicts of Interest: The authors declare no conflict of interest.

References

1. Deng, K.; Wan, G.; Deng, P.; Zhang, K.; Ding, S.; Wang, E.; Yan, M.; Huang, H.; Zhang, H.; Xu, Z.; et al. Experimental observation of topological Fermi arcs in type-II Weyl semimetal $MoTe_2$. *Nat. Phys.* **2016**, *12*, 1105–1110. [CrossRef]

2. Lv, B.Q.; Weng, H.M.; Fu, B.B.; Wang, X.P.; Miao, H.; Ma, J.; Richard, P.; Huang, X.C.; Zhao, L.X.; Chen, G.F.; et al. Discovery of Weyl semimetal TaAs. *Phys. Rev. X* **2015**, *5*, 031013. [CrossRef]

3. Xu, S.Y.; Alidoust, N.; Belopolski, I.; Yuan, Z.; Bian, G.; Chang, T.R.; Zheng, H.; Strocov, V.N.; Sanchez, D.S.; Chang, G.; et al. Discovery of a Weyl fermion state with Fermi arcs in niobium arsenide. *Nat. Phys.* **2015**, *11*, 748–754. [CrossRef]

4. Xu, S.Y.; Belopolski, I.; Alidoust, N.; Neupane, M.; Bian, G.; Zhang, C.; Sankar, R.; Chang, G.; Yuan, Z.; Lee, C.C.; et al. Discovery of a Weyl fermion semimetal and topological Fermi arcs. *Science* **2015**, *349*, 613–617. [CrossRef]

5. Xu, D.F.; Du, Y.P.; Wang, Z.; Li, Y.P.; Niu, X.H.; Yao, Q.; Dudin, P.; Xu, Z.A.; Wan, X.G.; Feng, D.L. Observation of Fermi Arcs in non-Centrosymmetric Weyl Semi-metal Candidate NbP. *Chin. Phys. Lett.* **2015**, *32*, 107101. [CrossRef]

6. Souma, S.; Wang, Z.; Kotaka, H.; Sato, T.; Nakayama, K.; Tanaka, Y.; Kimizuka, H.; Takahashi, T.; Yamauchi, K.; Oguchi, T.; et al. Direct Observation of Nonequivalent Fermi-Arc States of Opposite Surfaces in Noncentrosymmetric Weyl Semimetal NbP. *Phys. Rev. B* **2016**, *93*, 161112(R). [CrossRef]

7. Min, C.H.; Bentmann, H.; Neu, J.N.; Eck, P.; Moser, S.; Figgemeier, T.; Ünzelmann, M.; Kissner, K.; Lutz, P.; Koch, R.J.; et al. Orbital Fingerprint of Topological Fermi Arcs in the Weyl Semimetal TaP. *Phys. Rev. Lett.* **2019**, *122*, 116402. [CrossRef] [PubMed]

8. Niemann, A.C.; Gooth, J.; Wu, S.C.; Bäßler, S.; Sergelius, P.; Hühne, R.; Rellinghaus, B.; Shekhar, C.; Süß, V.; Schmidt, M.; et al. Chiral magnetoresistance in the Weyl semimetal NbP. *Sci. Rep.* **2017**, *7*, 43394. [CrossRef]

9. Huang, X.; Zhao, L.; Long, Y.; Wang, P.; Chen, D.; Yang, Z.; Liang, H.; Xue, M.; Weng, H.; Fang, Z.; et al. Observation of the chiral anomaly induced negative magneto-resistance in 3D Weyl semi-metal TaAs. *Phys. Rev. X* **2015**, *5*, 031023. [CrossRef]

10. Zhang, C.; Xu, S.Y.; Belopolski, I.; Yuan, Z.; Lin, Z.; Tong, B.; Bian, G.; Alidoust, N.; Lee, C.C.; Huang, S.M.; et al. Signatures of the Adler-Bell-Jackiw chiral anomaly in a Weyl Fermion semimetal. *Nat. Commun.* **2016**, *7*, 10735. [CrossRef]

11. Wang, J.; Wang, H.; Li, C.K.; Liu, H.; Yan, J.; Wang, J.; Liu, J.; Lin, Z.; Li, Y.; Wang, Y.; et al. Chiral anomaly and ultrahigh mobility in crystalline $HfTe_5$. *Phys. Rev. B* **2016**, *93*, 165127. [CrossRef]

12. Wang, J.; Li, H.; Chang, C.; He, K.; Lee, J.S.; Lu, H.; Sun, Y.; Ma, X.; Samarth, N.; Shen, S.; et al. Anomalous anisotropic magnetoresistance in topological insulator films. *Nano Res.* **2012**, *5*, 739–746. [CrossRef]

13. Dos Reis, R.D.; Ajeesh, M.O.; Kumar, N.; Arnold, F.; Shekhar, C.; Naumann, M.; Schmidt, M.; Nicklas, M.; Hassinger, E. On the search for the chiral anomaly in Weyl semimetals: The negative longitudinal magnetoresistance. *New J. Phys.* **2016**, *18*, 085006. [CrossRef]

14. Li, Y.; Wang, Z.; Li, P.; Yang, X.; Shen, Z.; Sheng, F.; Li, X.; Lu, Y.; Zheng, Y.; Xu, Z.A. Negative magnetoresistance in Weyl semimetals NbAs and NbP: Intrinsic chiral anomaly and extrinsic effects. *Front. Phys.* **2017**, *12*, 127205. [CrossRef]

15. Naumann, M.; Arnold, F.; Bachmann, M.D.; Modic, K.A.; Moll, P.J.W.; Süß, V.; Schmidt, M.; Hassinger, E. Orbital effect and weak localization physics in the longitudinal magnetoresistance of the Weyl semimetals NbP, NbAs, TaP and TaAs. *Phys. Rev. Mater.* **2020**, *4*, 034201. [CrossRef]

16. Chen, A.; Franz, M. Superconducting proximity effect and Majorana flat bands at the surface of a Weyl semimetal. *Phys. Rev. B* **2016**, *93*, 201105. [CrossRef]

17. Lu, B.; Yada, K.; Sato, M.; Tanaka, Y. Crossed surface flat bands of weyl semimetal superconductors. *Phys. Rev. Lett.* **2015**, *114*, 096804. [CrossRef]

18. Khanna, U.; Kundu, A.; Pradhan, S.; Rao, S. Proximity-induced superconductivity in Weyl semimetals. *Phys. Rev. B* **2014**, *90*, 195430. [CrossRef]

19. Li, Y.; Gu, Q.; Chen, C.; Zhang, J.; Liu, Q.; Hu, X.; Liu, J.; Liu, Y.; Ling, L.; Tian, M.; et al. Nontrivial superconductivity in topological $MoTe_{2-x}S_x$ crystals. *Proc. Natl. Acad. Sci. USA* **2018**, *115*, 9503–9508. [CrossRef]

20. Xing, Y.; Shao, Z.; Ge, J.; Wang, J.; Zhu, Z.; Liu, J.; Wang, Y.; Zhao, Z.; Yan, J.; Mandrus, D.; et al. Surface Superconductivity in the type II Weyl Semimetal $TaIrTe_4$. *Natl. Sci. Rev.* **2019**. [CrossRef]

21. Xu, S.Y.; Belopolski, I.; Sanchez, D.S.; Guo, C.; Chang, G.; Zhang, C.; Bian, G.; Yuan, Z.; Lu, H.; Feng, Y.; et al. Experimental discovery of a topological Weyl semimetal state in TaP. *Sci. Adv.* **2015**, *1*, e1501092. [CrossRef] [PubMed]

22. Huang, S.M.; Xu, S.Y.; Belopolski, I.; Lee, C.C.; Chang, G.; Wang, B.; Alidoust, N.; Bian, G.; Neupane, M.; Zhang, C.; et al. A Weyl Fermion semimetal with surface Fermi arcs in the transition metal monopnictide TaAs class. *Nat. Commun.* **2015**, *6*, 7373. [CrossRef] [PubMed]

23. Shekhar, C.; Nayak, A.K.; Sun, Y.; Schmidt, M.; Nicklas, M.; Leermakers, I.; Zeitler, U.; Skourski, Y.; Wosnitza, J.; Liu, Z.; et al. Extremely large magnetoresistance and ultrahigh mobility in the topological Weyl semimetal candidate NbP. *Nat. Phys.* **2015**, *11*, 645–649. [CrossRef]

24. Aggarwal, L.; Gayen, S.; Das, S.; Kumar, R.; Süß, V.; Shekhar, C.; Felser, C.; Sheet, G. Mesoscopic superconductivity and high spin polarization coexisting at metallic point contacts on the Weyl semimetal TaAs. *Nat. Commun.* **2017**, *8*, 13974. [CrossRef] [PubMed]

25. Wang, H.; Wang, H.; Chen, Y.; Luo, J.; Yuan, Z.; Liu, J.; Wang, Y.; Jia, S.; Liu, X.-J.; Wei, J.; et al. Discovery of tip induced unconventional superconductivity on Weyl semimetal. *Sci. Bull.* **2017**, *62*, 425–430. [CrossRef]

26. Wang, H.; He, Y.; Liu, Y.; Yuan, Z.; Jia, S.; Ma, L.; Liu, X.-J.; Wang, J. Ferromagnetic tip induced unconventional superconductivity in Weyl semimetal. *Sci. Bull.* **2020**, *65*, 21–26. [CrossRef]

27. Luo, J.; Li, Y.; Li, J.; Hashimoto, T.; Kawakami, T.; Lu, H.; Jia, S.; Sato, M.; Wang, J. Surface superconductivity on Weyl semimetal induced by nonmagnetic and ferromagnetic tips. *Phys. Rev. Mater.* **2019**, *3*, 124201. [CrossRef]

28. Li, Y.; Zhou, Y.; Guo, Z.; Chen, X.; Lu, P.; Wang, X.; An, C.; Zhou, Y.; Xing, J.; Du, G.; et al. Concurrence of superconductivity and structure transition in Weyl semimetal TaP under pressure. *NPJ Quantum Mater.* **2017**, *2*, 66. [CrossRef]

29. Kumar, P.; Sudesh, S.; Patnaik, S. Possible superconductivity in Weyl semimetal NbP. *AIP Conf. Proc.* **2016**, *1731*, 140063. [CrossRef]

30. Baenitz, M.; Schmidt, M.; Suess, V.; Felser, C.; Lüders, K. Superconductivity in Weyl Semimetal NbP: Bulk vs. Surface. *J. Phys. Conf. Ser.* **2019**, *1293*, 012002. [CrossRef]

31. Bachmann, M.D.; Nair, N.; Flicker, F.; Ilan, R.; Meng, T.; Ghimire, N.J.; Bauer, E.D.; Ronning, F.; Analytis, J.G.; Moll, P.J.W. Inducing superconductivity in Weyl semi-metal microstructures by selective ion sputtering. *Sci. Adv.* **2017**, *3*, e1602983. [CrossRef] [PubMed]

32. Arnold, F.; Shekhar, C.; Wu, S.C.; Sun, Y.; dos Reis, R.D.; Kumar, N.; Naumann, M.; Ajeesh, M.O.; Schmidt, M.; Grushin, A.G.; et al. Negative magnetoresistance without well-defined chirality in the Weyl semimetal TaP. *Nat. Commun.* **2016**, *7*, 11615. [CrossRef] [PubMed]

33. Besara, T.; Rhodes, D.A.; Chen, K.W.; Das, S.; Zhang, Q.R.; Sun, J.; Zeng, B.; Xin, Y.; Balicas, L.; Baumbach, R.E.; et al. Coexistence of Weyl Physics and Planar Defects in Semimetals TaP and TaAs. *Phys. Rev. B* **2016**, *93*, 245152. [CrossRef]

34. Du, J.; Wang, H.; Mao, Q.; Khan, R.; Xu, B.; Zhou, Y.; Zhang, Y.; Yang, J.; Chen, B.; Feng, C.; et al. Unsaturated both large positive and negative magnetoresistance in Weyl Semimetal TaP. *Sci. China* **2016**, *59*, 657406. [CrossRef]

35. Hu, J.; Liu, J.Y.; Graf, D.; Radmanesh, S.M.A.; Adams, D.J.; Chuang, A.; Wang, Y.; Chiorescu, I.; Wei, J.; Spinu, L.; et al. π Berry phase and Zeeman splitting of Weyl semimetal TaP. *Sci. Rep.* **2016**, *6*, 18674. [CrossRef]

36. Willerström, J.O. Stacking disorder in NbP, TaP, NbAs and TaAs. *J. Less-Common Met.* **1984**, *99*, 273–283. [CrossRef]

37. Xu, N.; Wang, Z.J.; Weber, A.P.; Magrez, A.; Bugnon, P.; Berger, H.; Matt, C.E.; Ma, J.Z.; Fu, B.B.; Lv, B.Q.; et al. Discovery of Weyl semimetal state violating Lorentz invariance in MoTe$_2$. *arXiv* **2016**, arXiv:1604.02116.

38. Jaroszynski, J.; Hunte, F.; Balicas, L.; Jo, Y.; Raičević, I.; Gurevich, A.; Larbalestier, D.; Balakirev, F.; Fang, L.; Cheng, P.; et al. Upper critical fields and thermally-activated transport of NdFeAsO$_{0.7}$F$_{0.3}$ single crystal. *Phys. Rev. B* **2008**, *78*, 174523. [CrossRef]

39. Santhanam, P.; Chi, C.C.; Wind, S.J.; Brady, M.J.; Bucchignano, J.J. Resistance anomaly near the superconducting transition temperature in short aluminum wires. *Phys. Rev. Lett.* **1991**, *66*, 2254. [CrossRef]

40. Nordström, A.; Rapp, Ö. Resistance-peak anomaly in metallic glasses: Dependence on currents and contact arrangement. *Phys. Rev. B* **1992**, *45*, 12577. [CrossRef]

41. Klimczuk, T.; Plackowski, T.; Sadowski, W.; Plebańczyk, M. A resistivity peak close to Tc in Nd$_{2-x}$Ce$_x$CuO$_{4-y}$ single crystals. *Phys. C Supercond.* **2003**, *387*, 203–207. [CrossRef]

42. Wang, J.; Singh, M.; Tian, M.; Kumar, N.; Liu, B.; Shi, C.; Jain, J.K.; Samarth, N.; Mallouk, T.E.; Chan, M.H.W. Interplay between superconductivity and ferromagnetism in crystalline nanowires. *Nat. Phys.* **2010**, *6*, 389–394. [CrossRef]

43. Vaglio, R.; Attanasio, C.; Maritato, L.; Ruosi, A. Explanation of the resistance-peak anomaly in nonhomogeneous superconductors. *Phys. Rev. B* **1993**, *47*, 15302. [CrossRef] [PubMed]

44. Chi, Z.; Chen, X.; An, C.; Yang, L.; Zhao, J.; Feng, Z.; Zhou, Y.; Zhou, Y. Pressure-induced superconductivity in MoP. *NPJ Quantum Mater.* **2018**, *3*, 28. [CrossRef]

45. Tinkham, M. Effect of fluxoid quantization on transition of superconducting films. *Phys. Rev.* **1963**, *129*, 2413. [CrossRef]

46. Tinkham, M. *Introduction to Superconductivity*, 2nd ed.; McGraw-Hill, Inc.: New York, NY, USA, 1996.

47. Yamafuji, K.; Kawashima, T.; Irie, F. On the angular dependence of the upper critical field in thin films. *Phys. Lett.* **1966**, *20*, 122–123. [CrossRef]

48. Yamafuji, K.; Kusayanagi, E.; Irie, F. On the angular dependence of the surface superconducting critical field. *Phys. Lett.* **1966**, *21*, 11–13. [CrossRef]

49. Zhang, H.M.; Sun, Y.; Li, W.; Peng, J.P.; Song, C.L.; Xing, Y.; Zhang, Q.; Guan, J.; Li, Z.; Zhao, Y.; et al. Detection of a superconducting phase in a two-atom layer of hexagonal Ga film grown on semiconducting GaN(0001). *Phys. Rev. Lett.* **2015**, *114*, 107003. [CrossRef]

50. Xing, Y.; Zhao, K.; Shan, P.; Zheng, F.; Zhang, Y.; Fu, H.; Liu, Y.; Tian, M.; Xi, C.; Liu, H.; et al. Ising Superconductivity and Quantum Phase Transition in Macro-Size Monolayer NbSe$_2$. *Nano Lett.* **2017**, *17*, 6802–6807. [CrossRef]

51. Lu, J.M.; Zheliuk, O.; Leermakers, I.; Yuan, N.F.Q.; Zeitler, U.; Law, K.T.; Ye, J.T. Evidence for two-dimensional Ising superconductivity in gated MoS$_2$. *Science* **2015**, *350*, 1353–1357. [CrossRef]

52. Ziegler, J.F.; Ziegler, M.D.; Biersack, J.P. SRIM—The stopping and range of ions in matter (2010). *Nucl. Instrum. Methods Phys. Res. B* **2010**, *268*, 1818–1823. [CrossRef]

53. Lin, X.; Zhu, Z.; Fauqué, B.; Behnia, K. Fermi surface of the most dilute superconductor. *Phys. Rev. X* **2013**, *3*, 021002. [CrossRef]

54. Yang, H.; Chen, G.; Zhu, X.; Xing, J.; Wen, H.H. BCS-like critical fluctuations with limited overlap of Cooper pairs in FeSe. *Phys. Rev. B* **2017**, *96*, 064501. [CrossRef]

55. Kasahara, S.; Watashige, T.; Hanaguri, T.; Kohsaka, Y.; Yamashita, T. Field-induced superconducting phase of FeSe in the BCS-BEC cross-over. *Proc. Natl. Acad. Sci. USA* **2014**, *111*, 16309–16313. [CrossRef] [PubMed]

56. Kasahara, S.; Yamashita, T.; Shi, A.; Kobayashi, R.; Shimoyama, Y.; Watashige, T.; Ishida, K.; Terashima, T.; Wolf, T.; Hardy, F.; et al. Giant superconducting fluctuations in the compensated semimetal FeSe at the BCS-BEC crossover. *Nat. Commun.* **2016**, *7*, 12843. [CrossRef] [PubMed]

57. Watashige, T.; Arsenijević, S.; Yamashita, T.; Terazawa, D.; Onishi, T.; Opherden, L.; Kasahara, S.; Tokiwa, Y.; Kasahara, Y.; Shibauchi, T.; et al. Quasiparticle excitations in the superconducting state of FeSe probed by thermal hall conductivity in the vicinity of the BCS-BEC crossover. *J. Phys. Soc. Jpn.* **2017**, *86*, 014707. [CrossRef]
58. Sheldrick, G.M. *CELL_NOW*; Georg-August-Universität Göttingen: Göttingen, Germany, 2008.

© 2020 by the authors. Licensee MDPI, Basel, Switzerland. This article is an open access article distributed under the terms and conditions of the Creative Commons Attribution (CC BY) license (http://creativecommons.org/licenses/by/4.0/).

Article

The Role of Chemical Substitutions on Bi-2212 Superconductors

Riccardo Cabassi [1,*], Davide Delmonte [1], Muna Mousa Abbas [2], Ali Razzak Abdulridha [3] and Edmondo Gilioli [1]

[1] Istituto dei Materiali per Elettronica e Magnetismo-CNR, Parco Area delle Scienze 37/A, 43124 Parma, Italy; davide.delmonte@imem.cnr.it (D.D.); edi.gilioli@gmail.com (E.G.)

[2] Department of Radiology Techniques, Al-Hadi University College, 10011 Baghdad, Iraq; muna_moussa@yahoo.com

[3] Department of Physics, College of Education for Pure Sciences, Babylon University, 11702 Babylon, Iraq; ali_rzzq@yahoo.com

* Correspondence: riccardo.cabassi@cnr.it

Received: 7 April 2020; Accepted: 22 May 2020; Published: 1 June 2020

Abstract: We present a study on the correlation of the superconducting critical temperature (Tc) and structural morphology with a chemically substituted high-temperature superconductor (HTS) (Bi,Pb)-2212 via Powder X-ray Diffraction (PXRD), Scanning Electron Microscopy (SEM), Energy Dispersive X-ray Spectroscopy (EDX), and dc magnetometry. The elements Zn, Y, Ti, and Nd are incorporated within the bismuth cuprate structure at amounts that extend the ranges currently found in literature.

Keywords: high-temperature superconductors; bismuth-based cuprates; Bi-2212

---

1. Introduction

Bismuth-based cuprates (BSCCO) were the first discovered high-temperature superconducting materials with no rare earth content [1]. Their crystal structure is formed by superconducting CuO_2 planes separated by charge reservoir layers crystallizing according to the general formula $Bi_2Sr_2Ca_{(n-1)}Cu_nO_{(2n+4+\delta)}$, where n is the number of CuO_2 planes in the unit cell. The superconducting phases occur at $n = 1$ (Bi-2201, $Tc = 10$ K), $n = 2$ (Bi-2212, $Tc = 85$ K), and $n = 3$ (Bi-2223, $Tc = 110$ K). While the highest Tc value pertains to Bi-2223, the most studied phase is Bi-2212 because of its higher thermodynamic stability and easier production route [2].

The superconducting properties of BSCCO systems—similarly to the other high-temperature superconductor (HTS) cuprates—depend on the charge-carrier concentration in the CuO_2 planes, which in turn is strongly affected by off-stoichiometricity. In particular, between oxygen excess and cationic disorder, the latter plays a dominant role [3]. With regard to this, it should be remarked that exact stoichiometries are extremely difficult to obtain in bulk compounds since, along with the starting composition, the sintering and processing conditions may affect the formation and amount of extra phases [4] or members of the homologous series with higher n values among the final results. Besides, an analysis of the ionic radii sizes shows that replacement of Ca,Sr with Pb,Cu is possible in presence of Cu^+ ions, with suitable δ values to preserve charge balance. Pb doping has been found long since [5] to yield favorable improvements: substituting Bi^{3+} with Pb^{2+} increases the formal Cu valency and thus the number of hole charge carriers [6], as a result, one has a sizable increase in the critical current $J_c(H)$ under magnetic field at optimal Pb content x = 0.16 with no significant variation in Tc [7]. For this reason, most of the currently produced BSCCO compounds belong to the Pb-doped family (Bi,Pb)-2212, and much research effort is aimed at finding other substitutions in search of further improvements.

In this study, we investigate the effect on the superconducting and structural/morphological properties of Zn, Y, Ti, and Nd substitutions in (Bi,Pb)-2212 up to high dopant concentrations. Whilst to the best of our knowledge, there are no studies reporting on the yttrium substitutions in (Bi,Pb)-2212, earlier studies on Zn-, Ti-, and Nd-substituted (Bi,Pb)-2212 are available for concentrations up to 0.04 [8], 0.15 [9], and 0.50 [10], respectively. In this work, we extend the investigated values to 0.32, 0.67, and 0.67.

2. Results

2.1. EDX

We analyzed (Bi,Pb)-2212 samples with increasing concentrations of Zn, Y Nd, Ti, and the undoped reference sample. The chemical compositions have been obtained from the Energy Dispersive X-ray Spectroscopy (EDX) spectra shown in Figure 1. The spectra were collected in the energy range 0–40 KeV, nonetheless, no interesting features appeared above 20 KeV. The elemental contents per formula unit are reported in Table 1. In the first row, the composition of the undoped parent compound, named X000, is reported and taken as a reference; while in the following rows, the doping element of each sample is specified in the 'Dopant' column. According to EDX analysis, the reference sample X000 has stoichiometry $Bi_{1.74}Pb_{0.39}Sr_{1.75}Ca_{0.73}Cu_{2.39}O_{7.87}$, therefore it belongs to the 2212 phase $Bi_{(2-y)}Pb_ySr_2CaCu_2O_{(8+\delta)}$, where y = 0.26 with some excess in Pb and Cu and deficiency in Sr and Ca. Assignment to the Bi-2212 phase is corroborated by the critical temperature Tc = 75 K resulting from the Zero Field Cooling (ZFC) magnetization curve that is repeated in each panel of Figure 4 for a convenient comparison with doped samples.

All the samples have similar Pb substitutions on the Bi site, apart from sample Nd05 however, which is not superconductive (see below). Moreover, the off-stoichiometry of the other cations is of the same type as for the sample X000, apart from the Ti-doped samples. On the basis of the available data, it is difficult to further speculate on the site location of each substitution.

Table 1. Formula unit elemental composition of the analyzed samples according to Energy Dispersive X-ray Spectroscopy (EDX) analysis and corresponding critical temperature Tc determined as the diamagnetic onset. The Tc value marked with '?' is of uncertain determination because of the low intensity of the diamagnetic signal. The 'Mass' column refers to the samples measured with the SQUID magnetometer.

Specimen	Mass (mg)	Bi	Pb	Sr	Ca	Cu	Dopant		Tc (K) 2212	2223
X000	32.4	1.74	0.39	1.75	0.73	2.39	-	-	79	
Zn01	114.6	1.68	0.38	1.70	0.74	2.42	Zn	0.09	75	
Zn05	34.6	1.65	0.35	1.62	0.81	2.25	Zn	0.32	71	100?
Y01	68.2	1.81	0.16	1.78	0.91	2.16	Y	0.18	102	
Y05	62.9	1.62	0.27	1.62	0.90	1.95	Y	0.64	79	
Ti01	79.8	1.86	0.30	1.69	0.99	2.07	Ti	0.08	65	108
Ti05	89.7	1.86	0.23	1.72	0.95	1.92	Ti	0.32	65	108
Ti10	72.2	1.63	0.27	1.75	0.81	1.88	Ti	0.67	65	104
Nd01	60.5	1.63	0.26	1.79	0.76	2.41	Nd	0.17	82	
Nd05	32.0	1.41	0.04	1.76	0.88	2.25	Nd	0.67	-	
Nd10	71.4	1.42	0.29	1.60	0.85	2.17	Nd	0.67	-	

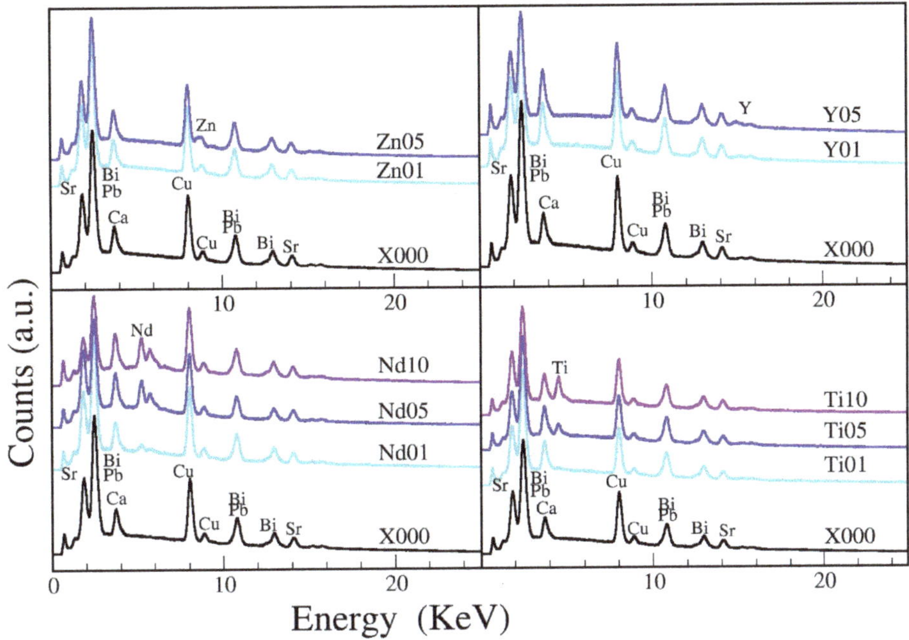

Figure 1. EDX patterns for the analyzed samples. From top to bottom: Y, Zn, Ti, and Nd substitutions. In each panel, the black line is the pattern of the reference undoped sample X000. An enlarged view of this figure, with detailed type of each emission line, is available as Figure S1 in Supplementary Materials.

2.2. PXRD

The quality of the products was also investigated by powder X-Ray diffraction (PXRD). The collected pattern of the reference sample X000 (black curve of Figure 2) matches the orthorhombic of the Bi-2212 phase (ICDD entry N.00-082-2278). A little amount of CuO is detected as the main spurious phase in all the patterns; this could explain the slight lack in composition detected for Cu by Energy Dispersive X-ray Spectroscopy (EDX) characterization. There are other traces of spurious peaks that are difficult to unambiguously address.

The Y and Zn series (see upper panels of Figure 2) do not affect the main phase stability, suggesting that Y and Zn easily substitute Ca—at least up to 0.5 in composition—within the structure. On the other hand, samples with the higher concentration of Nd and Ti ($x \simeq 0.5$–1) start to degrade the Bi-2212 phase, forming traces of spurious compounds (mixed Sr,Cu oxides). Interestingly, for Nd-substitution, the Bi-2212 phase is no longer superconducting for Nd $\geq$ 0.5, while in the case of Ti—in particular Ti10—the formation of higher member Bi-2223 phase was observed, in agreement with the double Tc onset detected (Figure 4c)—the latter in the temperature range 104–108 K.

The effect of a heavy Y- and Nd-doping clearly leads to a large contraction of the cell along the (0 0 10) c-axis (much larger than for Zn and Ti); reversely, the b-axis slightly enlarges, more significantly for Nd than for Y substitutions, as shown in the PXRD patterns zoomed in the 2θ region 27–34.5°, where the most intense and significative peaks are located for Y05, Nd05, and the reference sample X000 (Figure S3 in Supplementary Materials).

The simple one-to-one correlation between the crystallographic axes and the transition temperature Tc does not account for the complexity of the HTS cuprates. Since their discovery, a massive experimental and theoretical effort has been focused to find which interatomic distances within the unit cell affect the superconductivity, with particular attention to the CuO_2 planes aligned along the c-axis of the highly anisotropic HTS cell, where the superconductivity actually takes place. The number, distance, and interlayer coupling of the CuO_2 planes, their inner structure, *buckling angle*, in-plane cationic disorder, and the relation with the nonsuperconducting spacers were modified via cationic substitution. Despite the extensive work, the experimental results are still contradictory. In our case, the contraction of the c-axis of the heavily doped BSCCO samples is associated to the disappearance of the superconductivity in the Nd05 samples, and the drastic drop of the superconducting fraction in the Y05 one.

2.3. SEM Morphology

The surface morphologies observed by Scanning Electron Microscopy (SEM) images are reported in Figure 3. The reference sample features platelet-like grains of size in the order 10 μ, clearly depending on the synthesis/sintering procedures. The effect of doping on morphologies is of two opposite types, depending on the dopant element.

Moderate doping with Zn or Ti increases the size of platelets. Higher concentration of Ti, as in sample Ti10, induces a remarkably compact structure formed by merged layers and crossed by just few tubular cavities. Platelets' surface are clean, smooth, and the borders are neat.

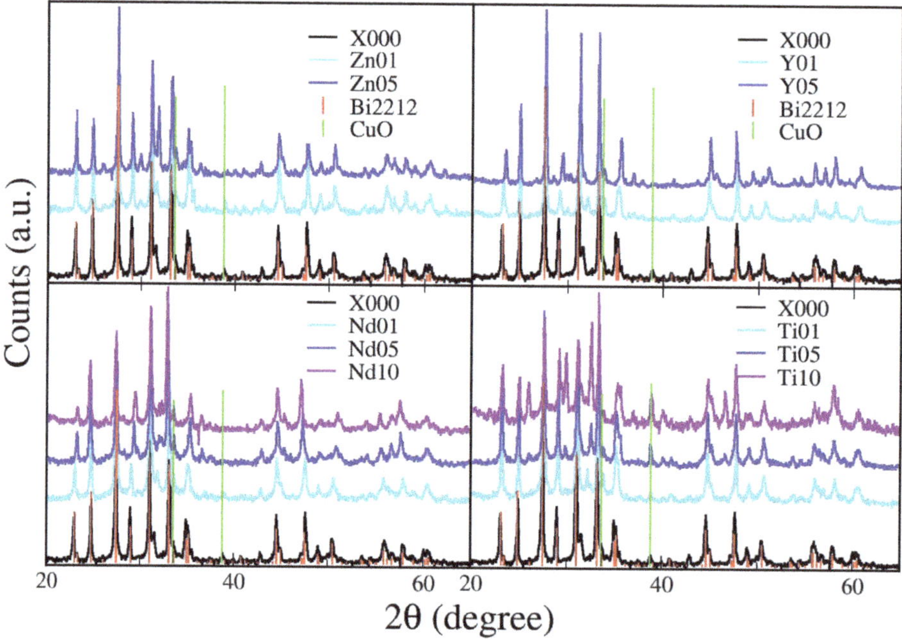

Figure 2. Powder X-ray Diffraction (PXRD) patterns for the analyzed samples. From top to bottom: Y, Zn, Ti, and Nd substitutions. In each panel, the black line is the pattern of the reference undoped sample. Red bars: Bi-2212 pattern from ICDD (chart n.00-082-2278). Green bars: CuO impurity pattern. An enlarged view of this figure is available as Figure S2 in Supplementary Materials.

On the contrary, samples with Y and Nd show smaller flat flakes in the order of $\sim$2 μm, with size decreasing on increasing the dopant content and approaching a granular aspect ratio for sample Y05. Similar trends in the evolution of morphology have already been observed with other dopant species [11].

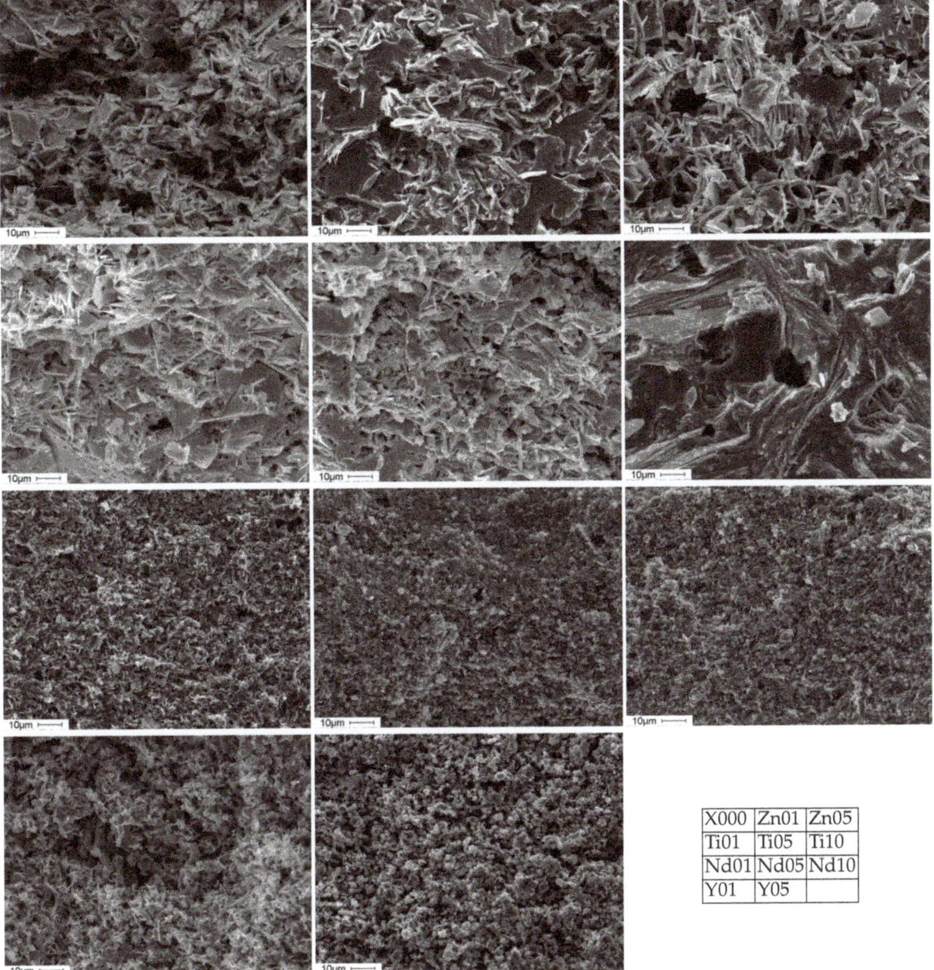

X000	Zn01	Zn05
Ti01	Ti05	Ti10
Nd01	Nd05	Nd10
Y01	Y05	

Figure 3. SEM surface micrographs of samples. From left to right, top to bottom: X000, Zn01, Zn05, Ti01, Ti05, Ti10, Nd01, Nd05, Nd10, Y01, Y05, as depicted in lower right corner.

2.4. Magnetic Characterization

In Figure 4, the resulting M(T) curves for reference and substituted samples are displayed. It is worth recalling that in a perfectly diamagnetic sample, the measured magnetic susceptibility is

$$\chi = \frac{-1}{4\pi(1-N)},\tag{1}$$

where N is the demagnetizing factor along the magnetization direction. It is easy to realize that N plays an important role on the absolute value of the measured signal in superconducting materials. For example, two cylinders with aspect ratio height/diameter equal to 1 and 2 have demagnetizing factors 0.232 and 0.069, respectively [12]: substituting into Equation (1), one obtains a difference between the measured susceptibilities amounting to a remarkable 30%. Considering that our samples had different and irregular shapes, a rigorous quantitative comparison of their diamagnetic intensities would be of little meaning. Therefore, the numerical value of diamagnetic intensities will be taken into account only in case of very notable differences between the samples, as in the case of Y substitutions. We will rather focus on estimating the critical temperatures Tc that, in the M(T) curves, can be identified with the temperature of the diamagnetic onset.

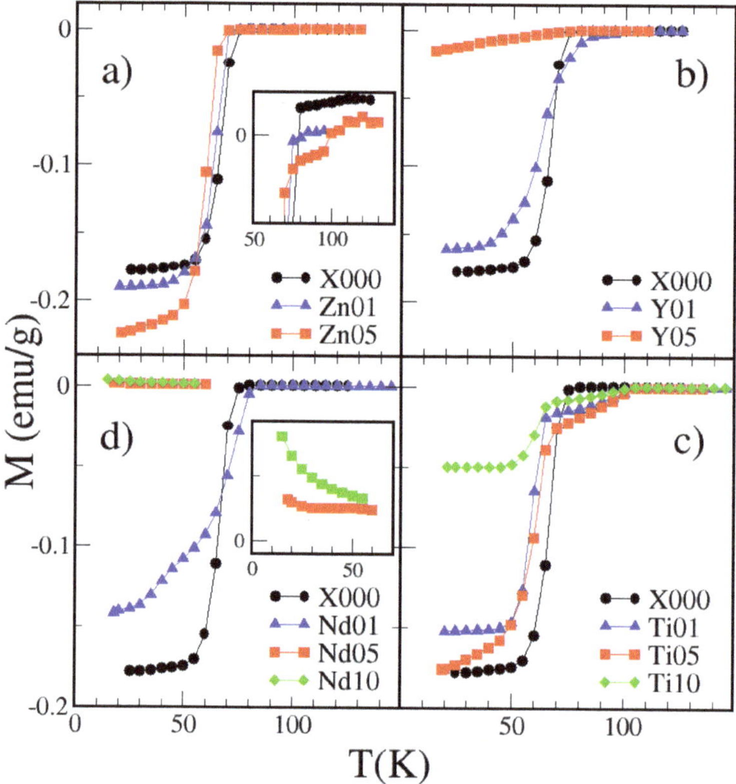

Figure 4. Zero Field Cooling (ZFC) magnetization curves of doped samples measured in applied magnetic field H = 100 Oe. X000 reference sample is also shown in each panel for comparison. Panels clockwise from top left: (**a**) Zn, (**b**) Y, (**c**) Ti, and (**d**) Nd substitutions. Inset of panel (**a**): magnified vision highlighting the diamagnetic signal above Tc of the Bi-2212 phase, suggesting the presence of a small fraction of Bi-2223 phase. Inset of panel (**d**): magnified vision of the paramagnetic signal of the nonsuperconductive samples Nd05 and Nd10.

2.4.1. Nd

In Nd-doped (Bi,Pb)-2212 a replacement of Nd^{3+} for Sr^{2+} ions occurs; measuring Nd occupations per f.u. x from 0.1 to 0.5 yields an optimal value $x = 0.2$ corresponding to the highest critical temperature $Tc = 94$ K. Above the optimal x value, Tc decreases down to $Tc = 82$ K for $x = 0.5$ [10]. Our Nd-doped samples comprise of a specimen with x close to the optimal value (Nd01, $x = 0.17$) and two overdoped samples with $x = 0.67$, thus extending the range analyzed in Reference [10]. As can be observed in panel (*a*) of Figure 4, sample Nd01 confirms an increase in Tc with respect to the undoped reference specimen X000; while in the two overdoped specimens, the superconductivity is completely suppressed. It is worth noting that Nd5 and Nd10 samples show the same Nd content; such a high Nd concentration seems to partially substitute Bi and Pb (Nd05) or Bi and Sr (Nd10); in both cases, the superconductivity vanishes.

2.4.2. Y

Yttrium substitution is among the earliest attempted to achieve a Tc increase in Bi-2212, finding the best doping value $x = 0.1\sim0.15$ with corresponding $Tc\sim$ 92 K [13,14]; this has been attributed to the effect of Y^{3+} substituting Ca^{2+}, which compensates for the hole excess in CuO_2 planes and restores the charge carrier concentration to its optimal value. As previously mentioned, in the literature, we did not find studies on Yttrium substitutions in (Bi,Pb)-2212, nonetheless, our measurements—panel (*b*) of Figure 4—show a similar effect on (Bi,Pb)-2212: sample Y01 ($x = 0.17$) confirms an increase in Tc of about 20 K with respect to the undoped reference specimen X000; on the other hand, increasing Y content up to $x = 0.64$, as in sample Y05, the diamagnetic onset is restored to the undoped values. Moreover, the diamagnetic intensity at the lowest measured temperature drops from -0.178 emu/g of the undoped sample to -0.015 emu/g only of Y05. This fall is too important to be ascribed to the demagnetizing effect discussed in Section 2.4, as Equation (1) would lead to unrealistic demagnetizing factors, and points rather to a strong reduction of the superconductive volume fraction of the specimen. The latter feature is likely related to the weak connectivity of the granular morphology evidenced in Table 3.

2.4.3. Zn

Zn substitution is known to induce a Tc depression in both Bi-2212 [15] and (Bi,Pb)-2212 [8]. Such depression has been explored in [8] for Zn content up to $x = 0.04$ and results to be in the order of $\Delta Tc\sim$ 10 K. We have extended the explored range up to $x = 0.32$, panel (*d*) of Figure 4, finding that the Tc depression maintains approximately the same magnitude even at such great doping content. A close inspection highlights the occurrence of diamagnetic signal even in the range above Tc of the Bi-2212 phase: this suggests the possible formation of Bi-2223 phase induced by a high level of Zn doping.

2.4.4. Ti

Ti doping in the BSCCO system yields the formation of Bi-2212/2223 multiphase samples with an improving of the superconducting properties for small Ti content x, and successive worsening for further x increasing up to the highest essayed value $x = 0.15$ [9,16]. The ionic radius of Ti^{4+}(68 pm) is similar to Cu^{2+}(72 pm), so it has been suggested that the former can substitute the latter in the crystal lattice. In a similar way, our Ti-doped samples feature the emerging of the high-Tc Bi-2223 phase at the expense of the Bi-2212 phase, panel (*c*) of Figure 4. According to literature, the nominal doping values of samples Ti01 and Ti05 should induce a sizable decrease on Tc in both the Bi-2223 and Bi-2212 components [16]. On the contrary, samples Ti01 and Ti05 have the same Tc values, while in sample Ti10, the variation of Tc can be considered negligible if one takes into account the very high x value.

3. Discussion and Conclusions

In the Nd- and Y-doped samples, our results confirm the worsening of the superconducting properties with x exceeding the values usually reported. For these samples, SEM micrographs show a reduced grain size, with deterioration of connectivity. On the contrary, Zn- and Ti-doped samples feature a substantial independence in the high x range. This could be at least partially related to the different grain sizes that appear to be greater than what can be found in literature. This is true for both Zn doping [8] and even more for Ti doping, where Tc appears to be independent on x. In principle, this could suggest that the majority of the Ti ions were segregated as impurities, rather than entering the BSCCO crystal lattice. However, SEM and EDX show a uniform compositional distribution, with no evidence of substituents segregation or inhomogeneities in platelet borders and surfaces. Moreover, by comparison with the images reported in Reference [9], one can notice the greater size of platelets, even from the low doping values. The latter, likely due to the different conditions of synthesis and sintering, can probably be the true origin of the different behaviors.

In short, the main conclusions are the following:

- We have measured (Bi,Pb)-2212 samples doped with Nd, Y, Zn, and Ti with content x exceeding the values reported in literature.
- Irrespective to the amount of substitutions, Ti induces the formation of the (Bi,Pb)-2223 phase in traces, clearly visible from the corresponding Tc value.
- The substituents show uniform compositional distribution, with no evidence of segregation, but leading to different grain size and grain connectivity.
- A high degree of substitutions (mainly for Y and Nd) induces similar structural changes (large contraction of the c-axis and a shorter expansion of b), but with a negative effect on Tc that remains unchanged for Y and has a strong reduction of the superconducting fraction; while it completely vanishes for Nd.

Owing to the structural and chemical complexity of the BSCCO HTSs and in spite of the long experimental activity, this work confirms the need of detailed studies to take into account the multiple structural changes induced by different chemical substitutions of (Bi,Pb)-2223 phase, and to clarify the controversial issue of the correlation with Tc.

4. Materials and Methods

Polycrystalline samples have been prepared as reported in Reference [17], starting from stoichiometric mixture of the high-purity binary oxides which were calcined in air at 800 °C for 24 h. The powder pressed into disk-shaped pellets using a manual hydraulic press type (SPECAC) under different pressures around 0.6 GPa. The pellets were sintered in air at 835~850 °C for 140 h. From the obtained pellets, fragments of irregular shape, and linear dimensions of $\sim$2 mm have been obtained and addressed to the various characterization techniques.

The phase identification was carried out by PXRD, using the Siemens D500 diffractometer, emitting $CuK\alpha_1$ and $CuK\alpha_2$ wavelengths (average $\lambda = 1.54178$ Å), with no filter for the $CuK\beta$.

The surface morphology of the samples was studied using SEM (Philips 515), operating at 25 kV, and equipped with an EDX "Phoenix" detector for compositional analysis. ZFC magnetization curves M(T) have been measured in applied magnetic field H = 100 Oe using a commercial SQUID magnetometer MPMS-5T (Quantum Design Co., San Diego, CA, USA).

Supplementary Materials: The following are available online at http://www.mdpi.com/2073-4352/10/6/462/s1, Figure S1. EDX patterns for the analysed samples, Figure S2. PXRD patterns for the analysed samples, Figure S3. Expanded PXRD patterns for Y05, Nd05 and reference sample X000.

Author Contributions: Conceptualization, R.C., D.D. and E.G.; methodology, R.C., D.D. and E.G.; validation, R.C., D.D. and E.G.; formal analysis, R.C., D.D. and E.G.; resources, R.C., D.D., M.M.A., A.R.A. and E.G.; investigation, R.C., D.D., M.M.A., A.R.A. and E.G.; writing—original draft preparation, R.C.; writing—review and editing, R.C., D.D. and E.G.; visualization, R.C. All authors have read and agreed to the published version of the manuscript.

Funding: This research received no external funding.

Acknowledgments: The authors thank P. Ferro (IMEM-CNR) for PXRD, F. Pattini (IMEM-CNR) for SEM data collection and S. Rampino (IMEM-CNR) for fruitful discussion.

Conflicts of Interest: The authors declare no conflict of interest.

Abbreviations

The following abbreviations are used in this manuscript:

BSCCO	Bismuth based cuprates
EDX	Energy Dispersive X-ray Spectroscopy
HTS	High Temperature Superconductor
PXRD	Powder X-ray Diffraction
SEM	Scanning Electron Microscopy
ZFC	Zero Field Cooling

References

1. Maeda, H.; Tanaka, Y.; Fukutomi, M.; Asano, T. A new high-Tc oxide superconductor without a rare earth element. *Jpn. J. Appl. Phys.* **1988**, *27*, L209. [CrossRef]
2. Manfredotti, C.; Truccato, M.; Rinaudo, G.; Allasia, D.; Volpe, P.; Benzi, P.; Agostino, A. Annealing temperature dependence of the 2223 phase volume fraction in the Bi–Sr–Ca–Cu–O system. *Phys. C Supercond.* **2001**, *353*, 184. [CrossRef]
3. Eisaki, H.; Kaneko, N.; Feng, D.L.; Damascelli, A.; Mang, P.K.; Shen, K.M.; Shen, Z.-X.; Greven, M. Effect of chemical inhomogeneity in bismuth-based copper oxide superconductors. *Phys. Rev. B* **2004**, *69*, 064512. [CrossRef]
4. Miao, H.; Kitaguchi H.; Kumakura H.; Togano K.; Hasegawa T.; Koizumi T. J_c enhancement of Bi-2212/Ag multilayer tapes by per-annealing and intermediate rolling process. *Adv. Cryogenic Eng.* **2000**, *46*, 559.
5. Togano, K.; Kumakura, H.; Maeda, H.; Yanagisawa, E.; Takahashi, K. Properties of Pb-doped Bi-Sr-Ca-Cu-O superconductors. *Appl. Phys. Lett.* **1988**, *53*, 1329. [CrossRef]
6. Coşkun, A.; Ekicibil, A.; Özçelik, B. Superconductivity of $Bi_{1.6}Pb_{0.4}Sr_2Ca_3Cu_4O_{12}$. *Chin. Phys. Lett.* **2002**, *19*, 1863.
7. Kumar, J.; Ahluwalia, P.K.; Kishan, H.; Awana, V.P.S. Significant Improvement in Superconductivity by Substituting Pb at Bi-site in $Bi_{2-x}Pb_xSr_2CaCu_2O_8$ with x = 0.0 to 0.40. *J. Supercond. Nov. Magn.* **2010**, *23*, 493–499. [CrossRef]
8. Bouaïcha, F.; Mosbah, M.F.; Amira, A.; Ait-Kaki, A.; Boussouf, N. Effect of Zn doping in Bi (Pb)-2212 superconducting ceramics. *Phys. Stat. Sol.* **2006**, *3*, 3036. [CrossRef]
9. Hamid, N.A.; Abd-Shukor, R. Effects of TiO_2 addition on the superconducting properties of Bi-Sr-Ca-Cu-O system. *J. Mater. Sci.* **2000**, *35*, 2325. [CrossRef]
10. Biju, A.; Syamaprasad, U.; Rao, A.; Xu, J.G.; Sivakumar, K.M.; Kuo, Y.K. Structural and transport properties of Nd doped (Bi, Pb)-2212. *Phys. C Supercond.* **2007**, *466*, 69. [CrossRef]
11. Bal, S.; Dogruer, M.; Yildirim, G.; Varilci, A.; Terzioglu, C.; Zalaoglu, Y. Role of Cerium Addition on Structural and Superconducting Properties of Bi-2212 System. *J. Supercond. Novel Magnet.* **2012**, *25*, 847. [CrossRef]
12. Chen, D.; Brug, J.A. Demagnetizing factors for cylinders. *IEEE Trans. Magnet.* **1991**, *21*, 3601. [CrossRef]
13. Zavaritskî, N.V.; Zavaritskî, V.N.; Mackenzie, A.P.; Orekhov, Y.F. Effect of yttrium on the increase in T_c of Bi(2212) crystals. *JETP Lett.* **1994**, *60*, 193.
14. Özçelik, B.; Nane, O.; Sotelo, A.; Madre, M.A. Effect of Yttrium substitution on superconductivity in Bi-2212 textured rods prepared by a LFZ technique. *Ceram. Int.* **2016**, *42*, 3418–3423. [CrossRef]
15. Kuo, Y.K.; Schneider, C.W.; Skove, M.J.; Nevitt, M.V.; Tessema, G.X.; McGee, J.J. Effect of magnetic and nonmagnetic impurities (Ni, Zn) substitution for Cu in $Bi_2(SrCa)_{2+n}(Cu_{1-x}M_x)_{1+n}O_y$ whiskers. *Phys. Rev. B* **1997**, *56*, 6201. [CrossRef]
16. Lu, Y.F.; Qiu, X.L.; Chao, X.X. The effect of titanium substitution in the $Bi_{1.6}Pb_{0.4}Sr_2Ca_2Cu_{3-x}Ti_xO_{10+y}$ system. *Solid State Commun.* **1995**, *95*, 259. [CrossRef]
17. Hermiz, G.Y.; Abbass, M.M.; Gilioli, E. Superconductivity of $(Bi_{0.7}Pb_{0.3})_2Ag_xSr_2Ca_2Cu_3O_{10+\delta}$ ($0 \leq x \leq 0.5$). *Atti Della Fondazione Giorgio Ronchi* **2009**, *64*, 1–8.

© 2020 by the authors. Licensee MDPI, Basel, Switzerland. This article is an open access article distributed under the terms and conditions of the Creative Commons Attribution (CC BY) license (http://creativecommons.org/licenses/by/4.0/).

MDPI

St. Alban-Anlage 66

4052 Basel

Switzerland

Tel. +41 61 683 77 34

Fax +41 61 302 89 18

www.mdpi.com

Crystals Editorial Office

E-mail: crystals@mdpi.com

www.mdpi.com/journal/crystals

www.ingramcontent.com/pod-product-compliance
Lightning Source LLC
LaVergne TN
LVHW070612170726
843515LV00004B/55